CHICAGO PUBLIC LIBRARY
HAROLD WASHINGTON LIBRARY CENTER
R0021947720

D1550375

DATE			

Environment and Energy

United Nations Economic Commission for Europe

Some other titles in this series from Pergamon Press

HOUSING FOR SPECIAL GROUPS

BEHAVIOUR OF WOOD PRODUCTS IN FIRE

PROTEIN AND NON-PROTEIN NITROGEN FOR RUMINANTS

FROZEN AND QUICK-FROZEN FOOD

FACTORS OF GROWTH AND INVESTMENT POLICIES

COAL: 1985 AND BEYOND

NON-WASTE TECHNOLOGY AND PRODUCTION

THE GAS INDUSTRY AND THE ENVIRONMENT

BUILDING RESEARCH POLICIES

STATISTICAL SERVICES IN TEN YEARS' TIME

HUMAN SETTLEMENTS AND ENERGY

ELECTRICAL LOAD-CURVE COVERAGE

ENGINEERING EQUIPMENT FOR FOUNDRIES

INTEGRATED STATISTICAL INFORMATION SYSTEMS 1977

FINE PARTICULATE POLLUTION

PROBLEMS OF THE AGRICULTURAL DEVELOPMENT OF LESS-FAVOURED AREAS

SELECTED WATER PROBLEMS IN ISLANDS AND COASTAL AREAS

Environment and Energy

Environmental Aspects of Energy Production and Use with particular reference to New Technologies A Report of the United Nations Economic Commission for Europe

Published for the

UNITED NATIONS

by

PERGAMON PRESS

OXFORD · NEW YORK · TORONTO · SYDNEY · PARIS · FRANKFURT

U.K.	Pergamon Press Ltd., Headington Hill Hall, Oxford OX3 0BW, England
U.S.A.	Pergamon Press Inc., Maxwell House, Fairview Park, Elmsford, New York 10523, U.S.A.
CANADA	Pergamon of Canada, Suite 104, 150 Consumers Road, Willowdale, Ontario M2J 1P9, Canada
AUSTRALIA	Pergamon Press (Aust.) Pty. Ltd., P.O. Box 544, Potts Point. N.S.W. 2011, Australia
FRANCE	Pergamon Press SARL, 24 rue des Ecoles, 75240 Paris, Cedex 05, France
FEDERAL REPUBLIC OF GERMANY	Pergamon Press GmbH, 6242 Kronberg-Taunus, Pferdstrasse 1, Federal Republic of Germany

First edition 1979

British Library Cataloguing in Publication Data

Economic Commission for Europe
Environment and energy.
1. Power resources - Environmental aspects
- Europe
2. Technological innovations - Europe
I. Title
621.4 TD195.E5 79-40550

ISBN 0-08-024468-8

In order to make this volume available as economically and as rapidly as possible the author's typescript has been reproduced in its original form.

Documents referred to by symbols (e.g. COAL/SEM.6/ Inf.1 and 2) are available from the UN documents distribution service.

Printed in Great Britain by Page Bros. (Norwich) Ltd., Norwich and London

Contents

Prefatory Note

This study has been prepared by the secretariat of the United Nations Economic Commission for Europe (ECE), pursuant to requests made jointly by the Senior Advisers to ECE Governments on Environmental Problems and the Senior Advisers to ECE Governments on Science and Technology at their respective third sessions in 1975, on the basis of a draft outline previously recommended by the Ad Hoc Meeting of Experts on Innovation in Energy Technology. At their fourth, fifth and sixth sessions, these two ECE bodies examined preliminary versions of this study; at their last session, the Senior Advisers on Environmental Problems decided that this project could be considered complete and invited Governments wishing to submit amendments to do so by 1 June 1978 in order that a final version of the document could be issued. Comments and additional information received from Governments and international agencies have been incorporated and, also, certain data have been updated.

To display each form of energy analysed, a double page format is used: the "Technological aspect" is described on the left-hand page and the "Environmental aspect" on the right. To further facilitate comparison between various forms, information has been classified under the same rubric for each of the 45 technologies studied.

Detailed tables of contents indicating the individual topics examined in Chapters 2, 3 and 4 are presented in the introductory pages of each chapter. For easy reference an index has also been provided.

Chapter One: Introduction

Brief review of existing sources of energy, their reserves and their present and future consumption patterns

1. In 1975, the world consumed approximately 8,000 million tons of coal equivalent (tce); 33 per cent in the form of hard coal, brown coal and peat, 44 per cent oil, 20 per cent natural gas and 3 per cent nuclear and hydro energy. Other forms of energy were either negligible (geothermal energy, tidal power, solar power) or not commercialized (wood, dung). World consumption is heavily concentrated in the ECE area which accounts for 75 per cent of world energy consumption.

2. The table (overleaf) gives the latest measured "recoverable" energy reserves. It is striking to note how incomplete and uncertain is our knowledge of the structure, location and volume of energy reserves. This uncertainty is mainly due to the fact that the exploration effort, its intensity and orientation, depend on the demand expected to materialize in ten or twenty years' time. The international comparability of data is hampered by differences in projected demand and demand patterns, differences in technical know-how and economic systems, in particular with regard to the allocation of costs. Accordingly, terms such as "reserves" and "resources" and definitions of what is "economically recoverable", while commonly used, contain an uncertain message. Such reservations emphasize the fact that these data provide only orders of magnitude, whose validity depends on assumptions made in the mid-1970s (or before) and which might be modified in the light of new technological and demand trends. It is thus worthwhile underlining that the figures show the proven measured reserves which are considered recoverable under the economic and technical conditions assumed to prevail in the next two or three decades.

3. From an environmental point of view, the data for the world show that the primary sources of energy creating the least environmental problems (solar power, natural gas, hydro power) either do not appear amongst the measured recoverable reserves, or are comparatively small. Direct 1/ solar power, although a very large source of energy, is not included in the status of "reserves" because its exploitation cannot yet be assessed. Natural gas and

---

1/ As opposed to indirect solar power such as hydro-electricity.

TABLE

Economically recoverable conventional forms of energy in the World and in ECE, 1976

	COAL	OIL	NATURAL GAS	TOTAL FOSSIL ENERGY	HYDRO-POWER a/	TOTAL CON-VENTIONAL ENERGY
	1	2	3	4	5	6
I. ABSOLUTE LEVEL (in 10^{18} joule)						
Europe	3 450	99	162	3 711	257	3 968
USSR	2 456	472	845	3 773	394	4 167
North America	5 780	294	340	6 414	562	6 976
ECE	11 686	865	1 347	13 898	1 213	15 111
WORLD	16 428	3 930	2 322	22 680	3 492	26 172
II. LOCATION (in % of world total)						
Europe	21	2.5	7	16	7	16
USSR	15	12	36	17	11	16
North America	35	7.5	15	28	16	27
ECE	71	22	58	61	34	59
WORLD	100	100	100	100	100	100
III. FUEL MIX (in % of total, conventional fuels)						
Europe	87	2	4	93	7	100
USSR	59	11	20	90	10	100
North America	83	4	5	92	8	100
ECE	77	6	9	92	8	100
WORLD	63	15	9	87	13	100
IV. DEPLETION TIMES AT 1965-1973 CONSUMPTION TRENDS (in years)						
Europe	167	2	7	29	renewable	...
USSR	78	15	24	33	"	...
North America	115	6	11	36	"	...
ECE	171	8	16	34	"	...
WORLD	101	17	20	35	"	...

Sources: World Energy Conference, Survey of Energy Resources 1976, London 1977, tables 1-3 (preface), for 1976; for hydro-power: WEC, Hydraulic Resources, London 1978.

a/ Total installed and installable capability x 100 years.

hydro power could count at the world level for approximately 12 per cent only of reserves. 1/ Thus, about 88 per cent of the world reserves considered technically and economically recoverable by the year 2000 are constituted of forms of energy which require more or less extensive treatment to reduce adverse effects on the environment.

4. The current prospects for an increased use of the comparatively clean conventional sources of energy are not bright. For example, a straightforward continuation of the historical (1965 to 1973) trends would prompt exhaustion of natural gas reserves in 20 years. In addition, hydro-electric energy sources harnessable at currently competitive costs are either extensively exploited or too far from consumption centres; moreover their exploitation, to an extent of 50 per cent, everywhere in the world (a clearly unrealistic assumption) would not cover more than 5 per cent of the projected world energy consumption in the year 2000 (compared with presently 3 per cent).

5. The pattern of reserves in ECE leaves hardly any ground for a policy of selective resource exploitation as the "clean" forms of energy (natural gas and hydro power) count for only 9 per cent of reserves: at historical consumption rates, natural gas deposits would be exhausted in about 16 years and economically exploitable hydro power sites are already harnessed to an extent of 60 per cent in Europe, 38 per cent in North America and 12 per cent in the USSR. Would increased reliance on imports of liquefied natural gas and of electricity (directly or in the form, for example, of hydrogen) from other regions provide a solution? The answer would seem to be that no significant relief could be expected from a global strategy of selective resource development: natural gas is comparatively scarce in the other regions and its liquefaction is expensive; the losses in the long-distance transmission of electricity (or the conversion losses if hydrogen is used as a vector) are still prohibitive. Thus, the ECE area seems more than any other region of the world compelled to comply with the environmentally unfavourable pattern of its energy reserves.

6. The consequences of discarding selective resource development as a means of diminishing the environmental disfunctions resulting from the use of energy at the source have to be fully recognized; continued reliance on fossil and increased importance of nuclear fuels, combined with energy demand, will increase the output of heat, CO_2 and SO_2. Larger releases of SO_2 and heat may have a significant impact on regional ecological balances and climates could be modified. Continued augmentation of the CO_2 content of the atmosphere might, in the course of a few generations, lead to changes in the global climate. Under these circumstances, the other options of controlling and reducing the environmentally harmful effects of energy consumption gain even greater importance:

- a consumer demand policy aimed at a slow-down of growth rates of energy consumption, for example through greater economy and efficiency in energy use; a study by the Executive Secretary estimates that between now and the early 1990s the overall efficiency in extraction, conversion, transmission and use of energy could be increased by a maximum of 100 per cent; the practically attainable improvement was assessed to be around one-third; in other words, economic growth rates could be "run" at growth rates of energy consumption which are one-third below present rates (E/ECE/883/Rev.1, table II.10).

1/ In order to make hydro power - a renewable form of energy - commensurable with non-renewable forms of energy, its annual potential output has been multiplied by 100.

- a continued generalization of pollution abatement techniques in all stages of energy production and use;

- a vigorous research and development effort in new environmentally-favourable forms of energy and technologies such as solar power, geothermal energy, hydrogen produced without thermal pollution from water in the presence of light radiation and an enzyme catalyst, fermentation of waste assisted by photosynthetic bacteria, sea currents, etc.

Chapter Two: Aspects of Present Technologies Applied to Some Energy Sources of Recognized Importance

(in particular the conversion into electricity of fossil fuels, nuclear and hydro-powers)

Contents

OPEN-CAST MINING OF SOLID FUELS

Technological aspect

1.1 Present technology

Principle: the over-burden is removed mechanically to uncover seams. In order to avoid flooding of the mine, regularization or diversion of water-courses is often necessary. Once mining has been completed, the site can be rehabilitated or developed for other uses.

Economic efficiency: greater than for underground mines (lower capital expenditure, higher technical efficiency, higher possibilities of adaptation and expansion and less time required to bring the mine into operation).

Importance in ECE region: 50 to 60 per cent of the coal produced in Europe comes from open-cast mines. There is little information regarding other solid fuels.

1.2 Future technology

Primarily, rationalization of production and development of means of reducing the impact on the environment.

2. Methods for reducing environmental impact

(i) Air pollution: debris may be covered with waste rock or with a layer of humus, which is then re-sown or re-timbered.

(ii) Water pollution: hydrogeological surveys taken before the commencement of mining operations, and careful planning of mining.

(iii) Rehabilitation or development of the site: the site may be filled with suitable materials - an operation which requires detailed studies - and returned to agriculture, forestry, etc. Large depressions may also be transformed into artificial water reservoirs for recreation or for urban, industrial or agricultural purposes (water supply, irrigation, etc.). Estimates made in the United States indicate that complete rehabilitation seems to increase the cost of producing coal by about $US 0.16 to 2.91 per ton (1977).

3. General assessment

This mining technology is likely to be widely used particularly for coal because of its economic efficiency (lower investment and operating costs than for underground mines) and also because of a revival of interest in coal for the production of electricity.

RELATED TOPICS

- *In situ* extraction of underground coal: pp. 42-43
- Tar sand exploitation: pp. 44-45
- Oil shale utilization: pp. 46-47
- Coal gasification: pp. 60-61
- Coal liquefaction: pp. 62-63

OPEN-CAST MINING OF SOLID FUELS

Environmental aspect

1. Air pollution

Exhaust gases from machines, oxidation products, dust raised by machines or by the wind. Tips containing a high percentage of inflammable substances are liable to spontaneous combustion. If the fire cannot be controlled in time, these tips can produce combustion gases for several decades.

2. Water pollution

Mainly caused by an accelerated erosion of the excavated material: pH changes, salts, particulates, etc.

3. Land use

Considerable. The area excavated is very seriously affected (total destruction of fauna and flora, and the surroundings are spoiled by tips, diversion of water courses, access roads, etc. ...).

4. Solid waste

Very large quantities (tips). Can be used for subsequent rehabilitation.

5. Noise

Fairly high during ripping of the over-burden and operations (caterpillar loaders, crushers, etc.). Detonations when explosives are used. Persons most seriously affected are the mine workers themselves.

6. Aesthetic aspect

Serious problem: excavated areas, tips, roads, etc. Can be eliminated, when mining has been completed, by rehabilitation of the site.

7. Others

None.

8. General assessment

Serious environmental effects. However do not present insoluble problems so long as the necessary precautions are taken (air and water pollution). When mining has been completed, the site can be rehabilitated or developed so as to eliminate, in the long term, all the land use and aesthetic problems.

GENERAL REFERENCES

Economic Commission for Europe, Coal Committee "Study of environmental problems resulting from coal industry activities", doc. COAL/R.11, 1974

J. Bastidon, "L'exploitation des carrières et son incidence sur l'environnement", Industrie minérale, August-September 1974

D.J. Davidson, "Open-cast Coal Mining. The Future and the Environment", Colliery Guardian, May 1975

C.W. Cook, "Surface-mine rehabilitation in the American West", Environmental Conservation, Vol. 3, No. 3 (1976), 179-183

N.P. Chironis, "Guide to Plants for Mine Spoils", Coal Age, Vol. 82, No. 7 (1977), 122-130

United Nations Environment Programme, "Draft Review of the environmental impact of production, transportation and use of fossil fuels" (1978), 132 pp.

OFFSHORE EXTRACTION OF OIL AND GAS

Technological aspect

1. Present technology

Methods: large platforms are built on the continental shelf, wells are drilled and the extracted oil or gas is transported to the seashore by pipelines or stored (for oil) in large floating reservoirs. Associate gases are either transported by pipelines, re-injected into the oil-bearing bed, processed at the platform (very expensive technology but desirable from the environmental point of view); flared (current practice but dangerous for platform equipment and creates air pollution problems) or burnt under the sea in special devices (new technology). Methods for increasing oil and gas production (pp. 38-39) can generally be applied.

Economic efficiency: lower economic efficiency compared to onshore oil fields resulting from a very complicated and expensive technology requiring rather large investments.

Importance in ECE region: offshore extraction of oil and gas has played an increasingly important role for the ECE countries because of the scarcity of onshore oil and gas deposits, potential further increase in world oil and gas prices and striving for self-sufficiency. For example, in 1972, western European countries covered only 4 per cent of their needs with local oil production. By means of the offshore oilfields of the North Sea it is hoped to raise this figure to 13 per cent in 1975 and about 20 per cent in 1980. Important gas and oilfields have been discovered along the northern shores of Alaska and Canada. Some are suspected in the Barents Sea and along the eastern coast of Greenland.

2. Methods of reducing environmental impact

There are no specific methods for offshore oilfields but the use of traditional methods, associated to safety measures is considerably more complicated. Methods have been developed in order to reduce the probability and the effects of potential oil spills at offshore platforms. Problems of the disposal of associate gas, drilling muds, brines and used water are very complicated from an environmental point of view. Generally muds, brines or used water should be re-injected into the oil-bearing formation.

3. General assessment

Offshore methods of oil and gas extraction have become a very important technology now that world oil recovery from the sea-bed has reached 20 per cent of total oil recovery. These methods will be widely used in the future.

RELATED TOPIC

- gas pipelines: pp. 12-13

OFFSHORE EXTRACTION OF OIL AND GAS

Environmental aspect

1. Air pollution

Mostly combustion products of flared associated gases and, in case of oil spill, evaporation of lighter hydrocarbon fractions.

2. Water pollution

Offshore production of gas should not create particularly significant problems when compared to offshore production of oil. Several major oil spills occurring at platforms have been reported and their effects on the marine environment and fisheries activities often extensively studied. Oil being biodegradable, rare and small-scale oil spills can be overcome through bacterial degradation. Large oil spills have a significant impact on marine flora and fauna and might affect the water cycle. If diluted in sea water, brines may have an ecological effect on surrounding marine communities.

Special attention and measures seem to be required during oil exploration in arctic and sub-arctic regions where the environment may be particularly sensitive and where, apparently, potential climatic disturbances might be induced by large-scale oil spills over and under ice by modifying the albedo.

3. Land use

Irrelevant.

4. Solid wastes

None to negligible.

5. Noise

No specific information but probably of minor importance except for staff working at the platform.

6. Aesthetic aspect

Fairly important problems due to derricks and harm to amenities, particularly beaches, in case of oil spill (tourism).

7. Others

Navigation safety; explosion and fire risks in case of important leaks.

8. General assessment

No acute problem with gas extraction. By contrast, oil extraction might lead to important environmental problems requiring special care and advanced technologies. Particular attention should be devoted to arctic and sub-arctic areas where environmental effects are still unclear and deserve further research.

GENERAL REFERENCES

United Nations Environment Programme "Review of the Impact of Production and Use of Energy on the Environment" (1976), document UNEP/GC/61/Add.1 and "Draft Review of the Environmental Impact of Production, Transportation and Use of Fossil Fuels" (1978).

W.J. Campbell and F.S. Martin: Oil and Ice in the Arctic Ocean: Possible Large-Scale Interaction, Science 181 (1973) 56-58; R.C. Ayers, H.O. Jahns and J.L. Glaeser: Oil Spills in the Arctic Ocean: Extent of Spreading and Possibility of Large-Scale Thermal Effects, Science 186 (1974) 843-845.

GAS PIPELINES

Technological aspect

1.1 Present technology

Pressurized (55-75 bars) gas is transported over long distances by pipes of a diameter up to 1,420 mm, mostly buried at about 1 m in depth, especially when crossing arable land. Sectionalizing valves are provided at about 30 km intervals and compressor stations (gas turbines) needed roughly every 100 km.

Economic efficiency: large capital investments required. For long distances and on the basis of equal heat energy transported, gas pipelines are economical when compared with electricity and solid fuel transportation, but more costly than liquid fuel transportation.

Importance in ECE region: in 1974 more than 98 per cent of the natural gas consumed was transported by pipeline. Long-distance gas pipelines will develop rapidly, especially in Europe where for the period 1969 to 1974 gas consumption has been growing at high annual rates: 12.6 per cent for Europe (excluding the USSR), 24.9 per cent for the European Economic Communities and 7.9 per cent for the USSR. Alaska, Northern Canada, Siberia, North Africa, North Sea and Middle East reserves would all require long-distance gas pipelines for bringing natural gas to large consumption centres.

1.2 Future technology

Pipes of larger diameters, higher pressures and lower temperatures will be used (a pressure rise from 55 bars to 75 bars increases through capacity by 30 to 35 per cent; at a given operating pressure, cooled natural gas (-65 to -70°C) doubles the transport capacity; liquefied natural gas at 40-45 bars increases capacity 3 to 4 times).

2. Methods for reducing environmental impact

There are no important environmental problems. By careful routing, cutting down trees and the risks of explosion can be reduced to a minimum.

3. General assessment

Natural gas transportation by pipeline will increase. Under present technical conditions, a gas pipe of about 900 mm diameter corresponds to the transportation of about 1,000 tons/hr of other fuels by roads, highways or inland waterways.

RELATED TOPICS

- offshore extraction of oil and gas: pp. 10-11
- coal gasification: pp. 60-61
- slurry pipelines: pp. 68-69
- capsule pipelines: pp. 70-71
- gas turbines: pp. 48-49

GAS PIPELINES

Environmental aspect

1. Air pollution

Very limited (compressors' exhausts if gas fired) during normal operation.

2. Water pollution

None during normal operation.

3. Land use

When the pipeline is buried, there are limited land use restrictions (proximity of buildings): these in general increase with the pressure of the system, 1/ the land can be returned to its former use as soon as construction is completed. 2/ Compressor stations, reception terminals and regulator stations are the only above-ground facilities. Land requirements are therefore very limited.

4. Solid waste

None.

5. Noise

Localized to regulating, metering and compressor stations (exhausts, air intake, engines and compressors, heat exchangers, blowdown). Compressor stations are generally located in rural areas and existing regulations specify that compressor noise level in residential areas should not exceed 35-40 db. 3/

6. Aesthetic aspects

Very limited if pipes can be buried. No trees on the right of way.

7. Others

Explosion risks in case of important and sudden leaks.

8. General assessment

No important environmental problems.

GENERAL REFERENCES

Economic Commission for Europe, Committee on Gas: "Report of the Symposium on the gas industry and environment", Minsk, 1977 (GAS/SEM.3/2).

United Nations Environment Programme, "Draft Review of the environmental impact of production, transportation and use of fossil fuels" (1978).

Hibberd, C., "The Environmental Impact of Natural Gas", Intern. J. Environmental Studies, Vol., 11, No. 2 (1977) pp. 99-112.

1/ In most countries, specific standards govern distances between pipelines and buildings.

2/ In Switzerland, planting trees is not allowed within 3 m on either side of a pipeline.

3/ In the USSR, the minimum distance between compressor stations and populated areas is 700 m at a maximum noise level of 80-90 db (at 2000 cycles per second).

OIL[1]-FIRED POWER PLANTS

Technological aspect

1.1 Present technology

Method of electricity production: the heat produced in the combustion chamber of the boiler vaporizes water and super-heats steam. The thermal energy is converted in the turbine into mechanical energy which is in turn converted into electric energy by a generator.

Thermal efficiency: 35-36 per cent seems near the maximum technically feasible.

Economic efficiency: depends on the fixed costs (initial investment) and variable costs (fuel, maintenance, labour and taxes). The cost of electric power produced in plants of this kind varies from country to country.

Importance in ECE region: in many European countries, oil-fired plants produce 60 per cent of the thermal production of electric power; this corresponds to about 40 per cent of the total electric power produced in Europe (excluding USSR) in 1972.

1.2 Future technology

Increased thermal conversion efficiency (see: new heat transfer media, pp. 50-51).

2. Methods for reducing environmental impact

(i) optimization of combustion so as to minimize the amounts of CO and NO produced;

(ii) desulphurization of the fuel oil and combustion gases; use of fuels with a low sulphur content;

(iii) partial utilization of the waste heat for district heating (combined production of heat and electric power); cooling towers. (But aesthetic impact; assisted draught towers seem to be a better answer.)

3. General assessment

Recent price increases for oil products in the world market have seriously affected the cost of the electricity produced in oil-fired plants. As a result, interest has shifted to other types of power station, in particular hydro-electric and nuclear power plants.

RELATED TOPICS

- offshore methods for oil and gas extraction: pp. 10-11
- other types of fossil-fuelled power plants: pp. 16-21
- nuclear power plants: pp. 22-31
- high voltage transmission lines: pp. 34-35
- new heat transfer media: pp. 50-51

1/ Standard fuel oil is composed of C:83.3 per cent, H:10.9 per cent, O:2.2 per cent and S:3.6 per cent, and produces about 40,000 kJ (9,600 kcal) per kg.

OIL-FIRED POWER PLANTS

Environmental aspect

1. Air pollution

Total emissions about 76,000 tons per annum for a modern 1,000 MWe plant burning 1.6 million tons of oil per year.

Aldehydes	0.2 kg/tce	Sulphur	13.5 kg/tce
CO	traces	Particulates	0.3 kg/tce
Hydrocarbons	0.3 kg/tce	Radioactivity	nil
NO_x	8.9 kg/tce		

2. Water pollution

Total: 3,000 to 6,000 tons per annum for a 1,000 MWe power plant

Thermal pollution: about 60 per cent of the energy consumed;
Biological fouling of condensers and cooling towers;
Chemical pollution: nil for the power plant itself except if biocides are used;
Radioactive pollution: nil.

3. Land use

About 4 km^2 for a 1,000 MWe power plant (without ancillary facilities).

4. Solid waste

Nil or negligible.

5. Noise

Not very serious (30 db at 100 m from the power plant); occasional whistling, audible up to a distance of 5 km, from the pressure regulators.

6. Aesthetic aspect

A serious problem due to chimney stacks, cooling towers and outside installations (storage tanks, railway lines, high-tension lines, etc. ...).

7. Others

Land use planning problems.

8. General assessment

The environmental problems raised by oil-fired power plants are fairly considerable from more than one point of view (in particular air pollution), but seem to be less serious than the problems inherent in coal-fire plants (see pp. 18-21).

GENERAL REFERENCES

Energy and the Environment: Electric Power, United States Government Printing Office, Council on Environment Quality, August 1973. Quoted in document ENV/R.11/Rev.1.

United Nations Environment Programme: "Review of the Impact of Production and Use of Energy on the Environment" (1976), document UNEP/GC/61/Add.1; and "Draft Review of the Environmental Impact of Production, Transportation and Use of Fossil Fuels" (1978) (data from National Academy of Engineering, Washington D.C.).

GAS-FIRED POWER PLANTS 1/

Technological aspect

1.1 Present technology

Method of electricity production: the heat produced in the combustion chamber of the boiler vaporizes water and super-heats steam. The thermal energy is converted in the turbine into mechanical energy, which is in turn converted into electric energy by a generator.

Thermal efficiency: 35-36 per cent seems near the maximum technically feasible (without gas turbines).

Economic efficiency: the market price of gas determines the extent to which it is used for electric power production. Although it is generally reserved as a basic product for industrial purposes, natural gas can, in producing countries, be competitive with other fuels in electric power production. For importing countries, the production of electric power from natural gas may prove to be uneconomical.

Importance in ECE region: accounts for between 2 and 8 per cent of the thermal production of electric power in the majority of ECE countries.

1.2 Future technology

See gas turbines (pp. 48-49).

2. Methods for reducing environmental impact

(i) optimization of combustion in order to minimize the amounts of CO and NO_x produced;

(ii) desulphurization of the fuel gas and combustion gases; use of fuels with a low sulphur content;

(iii) partial use of waste heat for district heating (combined production of electric power and heat); cooling towers (but aesthetic impact; assisted draught towers seem to be a better answer).

3. General assessment

Natural gas does not play a very important role in the production of electric power, and (barring the discovery of massive deposits) it is not likely to do so in future either. In the majority of cases, its use in thermal power plants has been decided upon not for purely economic reasons, but in order to protect the environment.

RELATED TOPICS

- offshore extraction of oil and gas: pp. 10-11
- gas pipelines: pp. 12-13
- other fossil-fuelled power plants: pp. 14-15 and 18-21
- nuclear power plants: pp. 22-31
- gas turbines: pp. 48-49
- coal gasification: pp. 60-61

1/ There are various types of gas (blast furnace gas, industrial gas, gas obtained from coal gasification (see pp. 60-61), petroleum gas, natural gas, etc. ...). The gas most widely-used in the electricity industry is natural gas, the usual composition of which is: CH_4 (methane): 94.0 per cent, C_2H_6 : 1.2 per cent, C_3H_8: 0.7 per cent, C_4H_{10}: 0.4 per cent, C_5H_{12}: 0.2 per cent, N_2: 3.3 per cent, CO: 0.2 per cent and SH_2 normally in slight traces. Its calorific value is about 35,800 kJ/m^3 (8,560 kcal/m^3) standard at 15°C (nm^3) and its specific weight is 0.765 kg/nm^3.

GAS-FIRED POWER PLANTS

Environmental aspect

1. Air pollution

Total emissions: about 24,000 tons per annum for a 1,000 MWe power plant

Aldehydes	0.06 kg/tce	Sulphur	0.1 - 0.2 kg/tce
CO	negligible	Particulates	0.9 kg/tce
Hydrocarbons	negligible	Radioactivity	nil
NO_x	20.9 kg/tce		

2. Water pollution

Total: About 1,000 tons per annum for a 1,000 MWe power plant

Thermal pollution: about 60 per cent of the energy consumed;
Biological fouling of condensers and cooling towers;
Chemical pollution: none from the power plant itself except if biocides are used;
Radioactive pollution: nil.

3. Land use

About 4 km^2 for a 1,000 MWe power plant.

4. Solid waste

None.

5. Noise

Not a very serious problem; occasional whistling, audible up to a distance of 5 km.

6. Aesthetic aspect

A fairly serious problem due to the chimney stacks, cooling towers and outside installations (high-tension lines, etc. ...).

7. Others

Installation of gas pipe-lines in urban areas (choice of emplacement, safety, etc. ...). Land use planning.

8. General assessment

Relatively limited environmental problems, except as regards waste heat and possible use of biocides. The environmental problems raised by gas-fired power plants, although not negligible, seem to be substantially less serious than those raised by coal (see pp. 18-21) or oil-fired plants (see pp. 14-15).

GENERAL REFERENCES

Energy and the Environment: Electric Power, United States Government Printing Office, Council of Environmental Quality, August 1973. Quoted in document ENV/R.11/Rev.1.
United Nations Environment Programme, "Review of the Impact of Production and Use of Energy on the Environment" (1976) document UNEP/GC/61/Add.1; and "Draft Review of the Environmental Impact of Production, Transportation and Use of Fossil Fuels" (1978) (data from National Academy of Engineering, Washington D.C.).
Economic Commission for Europe, "Proceedings of the Second Seminar on Desulphurization of Fuels and Combustion Gases", Washington D.C., (ENV/SEM.4/3).

HIGH CALORIE 1/ COAL FIRED POWER PLANTS

Technological aspect

1.1 Present technology

Method of electrical generation: the heat produced in the combustion chamber of the boiler vaporizes water and super-heats steam. The thermal energy is converted in the turbine into mechanical energy which in turn is converted into electric energy by a generator.

Thermal efficiency: 35-36 per cent seems near the maximum technically feasible.

Economic efficiency: depends on the price of coal in the countries considered. Coal can be competitive and will, in future, have a role to play in electric power production.

Importance in ECE region: varies; in most ECE countries, between 35 and 70 per cent of the thermal production of electricity results from this type of power plant.

1.2 Future technology

Fluidized bed combustion (pp. 56-57); new heat transfer media (pp. 50-51); development of combined cycles of higher efficiency using gas turbines (pp. 48-49).

2. Methods of reducing environmental impact

(i) optimization of combustion in order to minimize the amounts of CO and NO produced;
(ii) desulphurization of coal and combustion gases; use of fuels with a low sulphur content;
(iii) de-dusting of the smoke (possible up to 99.8 per cent);
(iv) ash tips raise environmental problem. The possibilities of using ashes in other economic sectors is under examination;
(v) partial use of waste heat for district heating; cooling towers (but for aesthetic impact, assisted draught towers seem to be a better answer).

3. General assessment

Coal now plays, and will continue to play, a very important role in the production of electricity. The number of ECE countries with large coal resources is limited; most import it. Protection of the environment and improved working conditions involve considerable expenditure, which affects the cost price of this source of energy. Also, high-calorie coal can be used with greater efficiency in other industrial sectors.

RELATED TOPICS

- open-cast mining of solid fuels: pp. 8-9
- other fossil-fuelled power plants: pp. 14-17 and 20-21
- nuclear power plants: pp. 22-31
- *in situ* extraction of underground coal: pp. 42-43
- new heat transfer media: pp. 50-51
- fluidized bed combustion: pp. 56-57
- coal gasification and liquefaction: pp. 60-63
- slurry and capsule pipelines: pp. 68-71

1/ Coal whose combustion releases 25,000 - 30,000 kJ/kg (6,000-7,000 kcal/kg) and would typically be composed of C: 67.9 per cent, H: 0.8 per cent, O: 4.8 per cent, N: 1.5 per cent, S: 0.5 per cent, H_2O: 9 per cent and ash 15.5 per cent.

HIGH CALORIE COAL-FIRED POWER PLANTS

Environmental aspect

1. Air pollution

Total emissions: about 165,000 tons per annum for a modern 1,000 MWe power plant

Aldehydes:	traces	Sulphur	14.7 kg/tce
CO	0.1 kg/tce	Particulates	2.0 kg/tce
Hydrocarbons	0.2 kg/tce	Radioactivity	traces
NO_x	9.1 kg/tce		

2. Water pollution

Thermal pollution: about 60 per cent of the energy consumed;
Biological fouling of condensers and cooling towers;
Chemical pollution: none, except if biocides are used;
Radioactive pollution: none.

3. Land use

About 4 km^2 for a 1,000 MWe power plant (storage and access facilities not included).

4. Solid waste

About 500,000 tons for operating a 1,000 MWe plant (slag).

5. Noise

Not very serious (30 db at 100 m from the power plant); occasional whistling, audible up to a distance of 5 km, from the pressure regulators. However, transport devices, such as cranes, conveyor belts, etc. could be noisy.

6. Aesthetic aspect

Problems due to the size of the installations, the height of the chimney stacks and cooling towers, and also the outside installations and ash tips. Often constructed in the industrial zones of cities.

7. Others

The choice of sites is largely determined by the availability of cooling water and the transportation of large quantities of coal. Normally such power plants are close to mine complexes when cooling water is available.

8. General assessment

The environmental problems raised by coal-fired power plants and their technological requirements (industrial infrastructure, land, corridors for high-tension lines, cooling, high chimney stacks, rail access, ash tips, etc. ...) are particularly serious from several points of view (air pollution, planning, etc.).

GENERAL REFERENCES

Energy and the Environment: Electric Power, United States Government Printing Office, Council of Environmental Quality, August 1973. Quoted in document ENV/R.11/Rev.1.
Economic Commission for Europe, "Proceedings of the Second Seminar on Desulphurization of Fuels and Combustion Gases", Washington D.C., (ENV/SEM.4/3).
United Nations Environment Programme, "Review of the Impact of Production and Use of Energy on the Environment" (1976) document UNEP/GC/61/Add.1; and "Draft Review of the Environmental Impact of Production, Transportation and Use of Fossil Fuels" (1978) (data from National Academy of Engineering, Washington D.C.).

BROWN COAL-FIRED POWER PLANTS

Technological aspect

1.1 Present technology

Method of electrical generation: since brown coal contains 20 to 40 per cent water and has an ash content varying from 8 to 40 per cent and a calorific value between 6,200 and 11,800 kJ/kg (1,500-2,900 kcal/kg), the power plant must have either special equipment for preparing the fuel before it is used in the power plant, or boilers specially designed to burn damp brown coal; this involves higher investment than for other thermal power plants. The heat produced vaporizes water and super-heats steam. The thermal energy is converted in the turbine into mechanical energy, which is in turn converted into electricity by a generator.

Thermal efficiency: can reach 35-36 per cent in modern plants.

Economic efficiency: although it requires higher specific investment, the cost of electric power produced from brown coal is fairly competitive, since brown coal (mostly from open-cast mines) is usually cheap.

Importance in ECE region: the following ECE countries are presently interested in brown coal: Bulgaria, Czechoslovakia, German Democratic Republic, Greece, Poland, Romania and Yugoslavia. In other countries, brown coal-fired plants - if there are any - are of limited interest for electric power production.

1.2 Future technology

No specific information available.

2. Methods for reducing environmental impact

(i) optimization of combustion in order to minimize the amounts of CO and NO produced;
(ii) desulphurization of fuel and combustion gases: use of fuels with a low sulphur content;
(iii) partial use of waste heat for district heating (combined production of electric power and heat); cooling towers (but aesthetic impact; assisted draught towers seem to be a better answer).

3. General assessment

Countries which have brown coal deposits, but no higher-grade energy resources, will increasingly rely on it for electricity production. At present, brown coal-fired power plants are economical but the environmental protection problems call for increasing attention, particularly with regard to solid waste, which might significantly reduce the economic efficiency of such power plants.

RELATED TOPICS

- open-cast mining of solid fuels: pp. 8-9
- other fossil-fuelled power plants: pp. 14-19
- nuclear power plants: pp. 22-31
- fluidized bed combustion: pp. 56-57

It should also be noted that the information given above, although relating specifically to brown coal-fired power plants, is also applicable to peat-fired plants which are very similar from the technical standpoint and show similar environmental problems.

BROWN COAL-FIRED POWER PLANTS

Environmental aspect

1. Air pollution

Very little specific information available; the amounts are probably of the same order of magnitude as in the case of coal-fired plants (see p.19).

2. Water pollution

Thermal pollution: about 60 per cent of the energy consumed;

Biological fouling: no specific information available; probably involves condensers and cooling towers;

Chemical pollution: varies from nil to minor amounts, except if biocides are used;

Radio-active pollution: no information - probably slight traces as in the case of coal-fired power plants (see p.19).

3. Land use

No specific information available; probably of the same order of magnitude as in the case of coal-fired power plants (see p.19).

4. Solid waste

Since the amount of ash and slag produced is very high, the problem of finding space to dump it is more acute than in the case of coal-fired power plants. Their possible utilization in other economic sectors should be regarded as a matter of priority.

5. Noise

Not very serious (30 db at 100 m from the plant); occasional whistling audible up to 5 km.

6. Aesthetic aspects

Serious problems because of the chimney stacks, cooling towers, high-tension lines, rail access and ash tips. In addition, problems inherent in open-cast mines (see pp. 8-9).

7. Others

To avoid the need to transport very large quantities of fuel, brown coal-fired thermal power plants are generally sited close to mining centres.

8. General assessment

Serious environmental problems deserving careful study by governments that might intend to use power stations of this kind on a large scale.

PRESSURIZED LIGHT-WATER REACTORS

Technological aspect

1.1 Present technology

Method of electrical generation: heat is produced by a controlled and maintained fission chain reaction using a fuel enriched in fissionable U-235. Ordinary (light) water is used as a coolant and at the same time acts as moderator. To achieve a high coolant outlet temperature without boiling, the system must be highly pressurized (about 160 kg/cm^2). A massive steel pressure vessel is used to contain the reactor core. The coolant is circulated through a number of primary loops containing steam generators and pumps. The resulting steam then drives turbo-generators.

Conversion efficiency: about 32 per cent.

Economic efficiency: demonstrated. Mainly depends on large investment costs. The fuel represents only a very limited fraction (in the order of 5 per cent) of the price of the produced electricity. The situation is evolving rapidly.

Importance in ECE region: in 1976 the installed capacity of ECE in PWR was 153,957 MWth in 49,474 MWe, representing about 60 per cent of the total nuclear installed capacity in the ECE countries.

1.2 Future technology

No dramatically new feature is to be expected. Present trends are toward providing increased safety and higher conversion efficiency by augmenting water temperature in the reactor vessel (see also new heat transfer media: pp. 50-51). Offshore plants.

2. Methods for reducing environmental impact

- contaminated air from nuclear installations is delayed before being released into the atmosphere in order to allow the decay of short-lived isotopes. Particularly harmful nuclides are trapped in filters or recovered.
- several methods for the management of highly radioactive wastes are envisaged. The most appropriate seems to be storage in salt deposits.
- partial recovery of waste heat for low grade energy utilization.
- increased overall efficiency by the combined production of electricity and heat for district heating and industrial purposes.
- increased safety measures.

3. General assessment

Investment costs and the availability of enriched fuel are probably the two main economic factors which will influence the future development of nuclear power. Pressurized light-water reactors are therefore sensitive to both issues. This type of reactor will however probably continue to represent the major share of the total nuclear installed capacity in the ECE region. It is furthermore increasingly recognized that problems of waste disposal or reprocessing will influence, in a decisive manner, future nuclear strategies.

RELATED TOPICS

- fossil-fuelled power plants: pp. 14-21
- other types of nuclear reactors: pp. 24-31
- new heat transfer media: pp. 50-51
- breeder reactors: pp. 58-59

PRESSURIZED LIGHT-WATER REACTORS

Environmental aspect

1. Air pollution
 Total: about 6,000 tons of materials and about 490,000 Ci of radioactive materials released yearly during the full fuel cycle (incl. mining, etc.) powering a 1,000 MWe station.
 Non-radioactive pollutants are mainly dust and particulates emitted during mining operations and exhaust gases from mining, transport, etc. equipment as well as some fluorine released during the enrichment step.
 Radioactive pollutants are mainly tritium, Kr^{85}, I^{129} and I^{131}. Amounts released at the power plant are low and vary notably with the cladding materials used (f. ex. 0.4 Ci of tritium/MWe per year with zircaloy and up to 17 Ci with stainless steel). Noble gases such as Kr^{85} are difficult to retain, while most of I^{129} and I^{131} is trapped in filters. The largest part of radioactivity is released at the reprocessing plant and is constituted of short-lived radionuclides which therefore do not accumulate in the atmosphere. On the other hand, tritium, Kr^{85} and I^{129} have longer half-lives and do accumulate in the environment.
2. Water pollution
 Total: about 21,000 tons of materials and 3,000 Ci of radioactive materials released yearly during the full fuel cycle for a 1,000 MWe power plant.
 Thermal pollution: about 2/3 of the gross thermal output. Cooling towers are commonly used. Significant heat discharges also occur during the enrichment step.
 biological fouling: mainly in condensers and cooling towers.
 Chemical pollution: some at the mining and refining stages. None at the power plant except if biocides are used.
 Radioactive pollution: slight to negligible at the power plant, moderate at the reprocessing plant. Major long-life nuclides released are tritium, Sr^{90}, Ru^{106} and Ce^{144}.
3. Land use
 About 77 km^2 for all operations (complete fuel cycle).
4. Solid waste
 About 2,600,000 tons released yearly during the full fuel cycle powering a 1,000 MWe station, mostly non-radioactive mining wastes and about 140 billion Ci of highly-compact radio-active wastes, the management of which is still at the research stage. Plutonium is recovered for use in breeder reactors (pp.58-59).
5. Noise
 Considered of minor importance. Comparable to fossil-fuelled power plant operations.
6. Aesthetic aspect
 Significant problem due to the size of installations, cooling towers, open air facilities, transformers, etc. Generally built in rural areas as a safety measure.
7. Others
 - Land use planning;
 - Safety concerns expressed by the public and part of the scientific community;
 - Non-proliferation questions;
 - Decommissioning.
8. General assessment
 The evaluation of the relative importance of some environmental factors listed above mainly depends nowadays on general attitudes towards economic policies; this is therefore not a matter answerable by the secretariat. From a purely technical point of view it can be said that a probably harmless low-level radioactive pollution is unavoidable during normal operation and that the problem of long-term disposal of highly radioactive wastes has still not found a solution which seems unequivocably satisfying from the economic, political, technical and environmental standpoints.

GENERAL REFERENCES: see p.25.

BOILING LIGHT WATER REACTORS

Technological aspect

1.1 Present technology

Method of electrical generation: heat is produced by a controlled and maintained fission chain reaction in a fuel enriched in fissionable U-235. Ordinary (light) water is used as a coolant and at the same time acts as moderator. Water is allowed to boil and produce steam at about half the system pressure of a PWR (about 70 kg/cm^2). This lower operating pressure allows a thinner walled but larger reactor vessel. Steam directly drives turbogenerators, and is then cooled in a condenser and recirculated through the reactor core.

Conversion efficiency: an average 33 per cent of the gross thermal output of BWR is converted into electricity.

Economic efficiency: comparable to PWR (see pp. 22-23).

Importance in ECE region: in 1976 the installed capacity of ECE in BWR was 79,784 MWth and 26,289 MWe, representing about 29 per cent of the total nuclear installed capacity in the ECE countries.

1.2 Future technology

Similar to PWR (see pp. 22-23).

2. Methods for reducing environmental impact

Similar to PWR (see pp. 22-23).

3. General assessment

Very comparable assessment as for PWR (see pp. 22-23). The share of BWR in the total installed nuclear capacity in the ECE region is not likely to change very significantly in the near future.

RELATED TOPICS

- fossil fuelled power plants: pp. 14-21
- other types of nuclear reactors: pp. 22-23 and 26-31
- new heat transfer media: pp. 50-51
- breeder reactors: pp. 58-59

BOILING LIGHT WATER REACTORS

Environmental aspect

1. Air pollution

Same remarks as for PWR (pp. 22-23). At the plant site, BWR are releasing less tritium but more noble gases (Kr^{85} and various isotopes of Xe).

2. Water pollution

Same remarks as for PWR. Radioactive pollution at the power plant site is virtually zero.

3. Land use

Similar to PWR.

4. Solid waste

Same remarks as for PWR.

5. Noise

Same comment as for PWR.

6. Aesthetic aspect

Comparable with PWR.

7. Others

Same features as for PWR.

8. General assessment

Comments made about PWR (see pp. 22-23) are also valid for BWR. It can in addition be stated that at the power plant site and in normal operation the radioactive contamination of the environment is lower than with PWR.

GENERAL REFERENCES

International Atomic Energy Agency (IAEA), Directory of nuclear reactors, Vol.X, Vienna 1976 (STI/PUB/397).
The IAEA has a number of technical publications. See IAEA Publications Catalogue 1976/77, Vienna 1976.
IAEA, Urban district heating using nuclear heat, Vienna 1977, document STI/PUB/461.
United States Council on Environment Quality, Energy and the Environment: Electric Power, United States Government Printing Office, 1973, as mentioned in ECE document ENV/R.11/Rev.1.
United Nations, Ionizing Radiations: Levels and Effects (United Nations Publications, sales Nos. E.72.IX.17 and 18) 2 vols.
United Nations "Sources and Effects of Ionizing Radiations" UNESCAR 1977 Report, (United Nations sales No.E.77.IX.1).

HEAVY WATER REACTORS

Technological aspect

1.1 Present technology

Method of electrical generation: various types of heavy water reactors have been developed. The most representative system operating on a commercial basis is the CANDU system: heat is produced by a controlled and maintained fission chain reaction in natural uranium fuel. Heavy water (D_2O) is used as coolant and acts as moderator. High pressures (about 100 kg/cm^2) are achieved as in PWRs (pp. 22-23) to prevent water from boiling. Steam produced in steam generators drives turbines. Other designs include boiling heavy water as for BWRs (pp. 24-25) and heavy water moderated, light water or gas cooled reactors (pp. 28-29).
Conversion efficiency: an average of 29 per cent of the gross thermal output of all designs of heavy water reactors is converted into electricity.
Economic efficiency: demonstrated. Large investment costs. The fuel represents a smaller fraction of the cost of produced electricity than with other types of reactors. Independent from enrichment facilities.
Importance in ECE region: in 1976 the installed capacity of the ECE in HWR was 9,847 MWth and 7,331 MWe, representing about 3 per cent of the total installed nuclear capacity in the ECE countries.

1.2 Future technology

As several types of HWR are presently under development or at the demonstration phase, new achievements can be expected notably in "mixed" systems such as heavy water moderated, light water or gas cooled reactors. The conversion efficiency will also probably be improved in the future.

2. Methods for reducing environmental impact

Similar as those mentioned for PWR (see pp. 22-23).

3. General assessment

Using natural uranium as fuel, heavy water reactors are independent from enrichment facilities making them attractive to some developing countries outside of the ECE region 1/ not planning on embarking in the enrichment work. Heavy water reactors will probably maintain a modest share in the total ECE installed nuclear capacity but are likely to represent a greater part of the installed capacity outside of the ECE region.

RELATED TOPICS

- fossil fuelled power plants: pp. 14-21
- other types of nuclear reactors: pp. 22-25 and 28-31
- new heat transfer media: pp. 50-51
- breeder reactors: pp. 58-59

1/ The International Atomic Energy Agency (IAEA) has carried out market surveys in this respect. See for example, IAEA, Power Reactors of Interest to Developing Countries, IAEA Bulletin, Vol. 17 No.3 (1975) pp. 36-42; or IAEA, Prospects for Utilization of Nuclear Power in Africa, *ibid*. (1976) Vol.18, No.1 pp. 40-44.

HEAVY WATER REACTORS

Environmental aspect

1. Air pollution

   Comparable to PWR (pp. 22-23); more tritium is released.

2. Water pollution

   Comparable to PWR. Higher thermal pollution resulting from lower conversion efficiency but no thermal pollution due to enrichment (although heavy-water "factories" consume large amounts of power).

3. Land use

   Similar to PWR.

4. Solid waste

   Probably comparable to PWR.

5. Noise

   Same comment as for PWR.

6. Aesthetic aspect

   Comparable to PWR with slightly-larger reactor building.

7. Others

   Same features as for PWR.

8. General assessment

   Comments made about PWR are also valid for HWR.

GENERAL REFERENCES

See p.25.

GAS-COOLED REACTORS

Technological aspect

1.1 Present technology

Method of electrical generation: heat is produced by a controlled and maintained fission chain reaction in natural uranium fuel. Graphite acts as moderator. GCR takes advantage of the fact that gases allow high temperatures at relatively moderate pressures (10-20 kg/cm^2 and 390°C) although their heat transfer capacity is quite limited. Carbon dioxide is generally used as coolant. Steam produced in steam generators drives turbines.

Conversion efficiency: an average 25 per cent of the gross thermal output of GCR is converted into electricity.

Economic efficiency: same remarks as for heavy-water reactors (pp. 26-27).

Importance in ECE region: in 1976, the installed capacity of the ECE in GCR was 29,493 MWth and 7,331 MWe, representing about 8 per cent of the total installed nuclear capacity in the ECE countries.

1.2 Future technology

Advanced gas-cooled reactors (AGR), achieving much higher efficiencies of conversion (about 40 per cent) by using for cooling CO_2 at higher temperatures and pressures (42 kg/cm^2 and 650°C) but fuelled with 2-3 per cent enriched uranium, are now at the demonstration stage at the industrial scale. If they prove efficient and reliable in the mid-term, AGR are likely to take over a significant part of the reactors market. See also high-temperature gas-cooled reactor (pp. 28-29).

2. Methods for reducing environmental impact

Similar to those mentioned for PWR (pp. 22-23). AGR represent a considerable improvement over GCR from the thermal pollution viewpoint.

3. General assessment

Using natural uranium as fuel GCR are independent from enrichment facilities, an advantage which is lost in the case of AGR. However, AGR will probably progressively take over GCR. Both types of reactor are also attractive from the economic viewpoint as no shut down is normally required for refuelling.

RELATED TOPICS

- fossil fuelled power plants: pp. 14-21
- other types of nuclear reactors: pp. 22-27 and 30-31
- new heat transfer media: pp. 50-51
- breeder reactors: pp. 58-59

GAS-COOLED REACTORS

Environmental aspect

1. Air pollution

   Comparable to PWR (pp. 22-23).

2. Water pollution

   Comparable to but slightly higher than PWR. Increased thermal pollution resulting from lower conversion efficiency in GCR, but no thermal pollution due to enrichment, except for AGR (see also p. 27).

3. Land use

   Similar to PWR.

4. Solid waste

   Probably comparable to PWR.

5. Noise

   Same comment as for PWR.

6. Aesthetic aspect

   Comparable to PWR.

7. Others

   Same features as for PWR.

8. General assessment

   Comments made for PWR (pp. 22-23) are also valid for GCR and AGR.

GENERAL REFERENCES

See p. 25.

HIGH TEMPERATURE GAS-COOLED REACTORS

Technological aspect

1.1 Present technology

Method of electrical generation: heat is produced by a controlled and maintained fission chain reaction in 4-5 per cent enriched uranium fuel. Graphite acts as moderator and helium gas at high pressure and temperature ($\sim$ 50 kg/cm^2 and 780°C) transfers the heat produced to steam generators. Considerable interest has been shown in utilizing processed heat from HTGR for some large scale industrial processes requiring high temperatures such as coal gasification and steelmaking.

Conversion efficiency: about 40 per cent.

Economic efficiency: in 1976 the few HTGR in exploitation were still at the industrial demonstration stage.

Importance in ECE region: only two commercial-size HTGR in the ECE region in 1976 representing 630 MWe of installed capacity (less than 1 per cent of the total installed capacity in the ECE region).

1.2 Future technology

HTGR fuelled with highly enriched uranium (80-90 per cent U-235) mixed with thorium would convert it during operation to uranium-233 which is a highly-efficient nuclear fuel and can be recovered during reprocessing of the fuel. This type of reactor would be entirely dependent on enrichment or reprocessing facilities.

2. Methods for reducing environmental impact

No specific information available. Probably fairly similar to those mentioned for PWR (pp. 22-23).

3. General assessment

A higher conversion efficiency and a wider spectrum of potential applications render HTGR quite attractive from an economic point of view. However, their viability and their economic efficiency have still to be demonstrated.

RELATED TOPICS

- fossil fuelled power plants: pp. 14-21
- other types of nuclear reactors: pp. 22-29
- new heat transfer media: pp. 50-51
- breeder reactors: pp. 58-59

HIGH TEMPERATURE GAS-COOLED REACTORS

Environmental aspect

1. Air pollution

   No information available on commercial-size facilities.

2. Water pollution

   ibid.

3. Land use

   Probably comparable to PWR (pp. 22-23).

4. Solid waste

   No information available.

5. Noise

   No information available.

6. Aesthetic aspect

   Comparable to PWR.

7. Others

   Probably same features as for PWR (pp. 22-23). It should be noted that 80-90 per cent enriched uranium as well as probably uranium-233 (see 1.2 opposite) represents so-called "weapon-grade fuels".

8. General assessment

   Made difficult by the absence of appropriate data. Probably quite similar to that made for PWR (pp. 22-23).

GENERAL REFERENCES

See p. 25.

HYDRO POWER PLANTS

Technological aspect

1.1 Present technology

Method: dams retain water creating a reservoir and a fall (kinetic energy) which activates turbines producing electricity. Dams situated in mountainous areas, such as the Alps, create high pressure falls producing large quantities of electricity with relatively small amounts of water. This type of power plant is very flexible and is often used as an energy storage capacity for meeting peak demands. On the other hand, dams situated across rivers utilize large amounts of flowing water for producing electricity and are particularly suitable for meeting basic load needs.

Efficiency of conversion: 75 to 95 per cent.

Economic efficiency: demonstrated. Large investment but no fuel costs. Often associated with flood control, watercourses regulation and irrigation.

Importance in ECE area: about 200,000 MW of total installed capacity. In Europe, 133,000 MWe of installed capacity in 1970, representing 26 per cent of the over-all electricity production. The planned installed capacity in 1985 is 240,000 MWe representing 15 per cent of the total electricity supply. Pumped storage plants equipped with reversible turbines play an increasingly important role as large energy storage facilities.

1.2 Future technologies

(a) very large facilities flooding huge areas, notably in sub-arctic regions (e.g. Canada)

(b) "micro-hydro" facilities for meeting local needs

(c) more exotic schemes such as the use of arctic glaciers (Greenland) and naturally occurring depressions (helio-hydro-power plants)

2. Methods for reducing environmental impact

Dams which stop fish migration paths have special canals with "stairs" which allow fish to progressively jump over the obstacle. No other method seems applicable.

3. General assessment

Hydropower is one of the oldest energy technologies ever developed and has played an important role in the development of the ECE region. It has been estimated that, in the whole of the ECE region, 25 per cent of the economically exploitable hydropower potential was harnessed at the end of 1971. This renewable source of energy could therefore continue to play an important role in meeting the electricity needs of ECE.

RELATED TOPICS

- other types of power plants: pp. 14-31
- tidal energy: pp. 94-95

HYDRO POWER PLANTS

Environmental aspect

1. Air pollution

   None, except during the construction phase (dust, exhaust gas, etc.).

2. Water pollution

   Decreased auto-epuration capacity of the water body. Proliferation of anaerobic bacteria in water bodies which are already polluted.

3. Land use

   May be considerable (e.g. 6,000 km^2 for the 12,000 MWe Lagrande complex in Canada).

4. Solid waste

   None, except during the construction phase

5. Noise

   None

6. Aesthetic aspect

   Varies considerably from site to site. Potential fish killing when passing through turbines. Diversion of watercourses. Sometimes contributes to the improvement of touristic potentialities of the areas concerned.

7. Others

   Hazards from dam failure. Accidental oil releases from turbines and transformers. Barriers to fish migration. Social impact when communities or large territories are flooded. Potential local climatic changes (wind pattern and speed, increased evaporation, change in ground albedo). Potential seismic effect due to the weight of water. Silting.

8. General assessment

   Small scale application of hydropower creates only marginal environmental problems and seems preferable to large scale schemes. Can be considered as environmentally compatible.

GENERAL REFERENCES

Economic Commission for Europe, Long-term prospects of the electric power industry in Europe 1970-1985, (ECE/EP/7), New York, 1974, United Nations publication, Sales No. E.74.II.E/Mim.5.

Economic Commission for Europe Increased Energy Economy and Efficiency in the ECE region (E/ECE/883/Rev.1), New York, 1976, United Nations publication, Sales No. E.76.II.E.2.

International Institute for Applied System Analysis "Power from Glaciers: the hydropower potential of Greenland's glacial waters", Report RR-77-20, 1977.

HIGH VOLTAGE TRANSMISSION LINES

Technological aspect

1. Present technology

Method: electricity is transported through overhead cables at high voltages sometimes over long distances. Power plants are usually interconnected through networks of power transmission lines which allows flexibility in generating plant location and operation.

Economic efficiency: a 200 mile long (322 km), 345 kV line transmits power with more than 98 per cent efficiency.

Importance in ECE region: most of the power plants of ECE are interconnected by high voltage transmission lines.

1.2 Future technology

Three new technologies for underground power transmission are being studied in addition to ways of improving conventional overhead lines:
(a) transmission cables insulated with compressed gas; (b) cryogenic transmission lines; (c) super-conducting transmission lines.

However, underground cables now available often have too low a transmission capacity to replace overhead lines. Underground lines are limited by the inability of the ground to absorb heat produced in the cable but special sand fills and/or water cooling overcomes this problem.

2. Methods for reducing environmental impact

Underground high voltage transmission lines would reduce unfavourable aesthetic aspects although in some cases, underground lines seem to sterilize more land than when using towers and overhead cables.

3. General assessment

Overhead transmission lines are reliable, easy to repair and efficient.

Underground transmission may replace overhead lines in certain situations where overhead transmission is impossible (offshore power plant) or unsafe, such as at intersections of lines with super-highways, airport runways or in urban areas, or in situations where the visual environment deserves such a protective measure.

RELATED TOPICS

- various types of power plants: pp. 14-33 (fossil, nuclear and hydro); pp. 58-59 (breeder reactors); pp. 74-75 (geothermal); pp. 82-83 (solar); pp. 90-91 (fusion); pp. 94-95 (tidal); and pp. 98-99 (ocean temperature gradient).

HIGH VOLTAGE TRANSMISSION LINES

Environmental aspect

1. Air pollution

   None

2. Water pollution

   If used as isolating material, PCBs could accidentally be released from transformers at both ends of the transmission line.

3. Land use

   Considered as a major problem, as a significant amount of land is required. Overhead transmission lines typically require 12 acres per mile ($30,000\ m^2/km$).

4. Solid wastes

   None

5. Noise

   Slight whistling

6. Aesthetic aspect

   Can be serious (unsightly).

7. Others

   Safety concerns for human beings and large birds.

8. General assessment

   The public reaction to the presence and appearance of overhead high tension lines is an important factor. This seems to be the major obstacle to the use of such facilities, although concerns have also been expressed with respect to human health and the protection of wild fauna (notably eagles and owls).

GENERAL REFERENCES

A. Hammond, W. Metz and T. Maugh II, Energy and the Future, American Association for the Advancement of Science, Washington, D.C., (1973), p.102.

U. Morosoff "People's safety from the influence of electric fields generated by high voltage transmission lines" Review of scientific work of the USSR Trade Unions Institute for Labour Safety. Profizdat Vol. 87 (1972) pp. 23-27.

Economic Commission for Europe, Committee on Electric Power, Electricity and Environment, (ECE/EP/22), volume I, 1977.

J.R. Allen, L.A. Carstens and D.H. Norback "Biological effects of the polychlorinated biphenyls in non-human primates" Proceedings of the International Symposium on Recent Advances in the Assessment of the Health Effects of Environmental Pollution, Paris, 24-28 June 1974, pp. 385-396.

Chapter Three: Techniques Under Development Which Relate to Energy Sources of Recognized Importance

Contents

METHODS OF INCREASING GAS OR OIL EXTRACTION EFFICIENCY

Technological aspect

1. Present technology

Methods: (i) Injection of water: the formation is maintained under pressure by pumping water into it. Depending on the type of oil, this increases oil recovery up to 50 per cent as compared to 20-30 per cent by primary methods. This method is not effective for high viscosity oils.

(ii) Additives : chemicals are added to increase the effectiveness of the pumped water method. Depending on the characteristics of the oil, geological formation, penetrability of oil-bearing bed and some other factors, the following water additives are used: surfactants, polymers, ammonia, carbon dioxide, colloid of silicon oxide, etc. These additives are usually used in combination with other techniques (edging technique, changes of water pressure and régime, selection of number and place of input and output wells, etc.) and allow an increase in oil recovery from 50-55 per cent to 65-70 per cent.

(iii) Injection of high pressure air or gas (up to 640 atm.)

(iv) Thermal methods: such as, for example, injection of steam, in situ combustion of oil, etc. (heat effect).

(v) Shaft methods: allow increased oil extraction to 50-60 per cent, and, in combination with thermal methods, up to 90 per cent.

(vi) Combinations of these methods.

The same methods are used to increase gas and gas condensate outputs.

Economic efficiency: Most of the methods listed above (except shaft method) have a high economic efficiency: they increase the productivity of wells and the total oil output. The shaft method is less efficient.

Importance in ECE region: These methods appear to be widely used in ECE oil-producing countries.

1.2 Future technology

Underground nuclear explosions (see pp. 40-41).

2. Methods for reducing environmental impact

Depending on the technology used to enhance extraction, a great number of protection measures can be taken to eliminate air and water pollution. These include various methods of water and solvents refining and recycling, refinery systems of associated gas and systems of associated gas injection into wells.

3. General assessment

Since the primary methods of oil extraction allow the recovery of only 20-30 per cent of an oil deposit, and taking into account the scarcity of oil reserves, all methods of increasing oil output will continue to be further developed and widely used in oil-producing countries of the ECE region.

RELATED TOPICS

- offshore extraction of gas and oil: pp. 10-11
- underground nuclear explosions: pp. 40-41
- tar sands exploitation: pp. 44-45
- oil shale utilization: pp. 46-47

METHODS OF INCREASING GAS OR OIL EXTRACTION EFFICIENCY

Environmental aspect

1. Air pollution

   No specific information available. Should be of minor importance and varying with the technique used.

2. Water pollution

   Important chemical pollution in case safety measures would fail when chemicals such as surfactants, ammonia or silicone derivatives would be used in the vicinity of a water body. Used water should be refined and recycled, or re-injected into the geological formation.

3. Land use

   No specific information available; should be minor.

4. Solid wastes

   Minor problem (containers, etc.)

5. Noise

   No specific information; probably minor problems except for workers.

6. Aesthetic aspect

   No specific information; probably no serious problems additional to those associated with regular oil drilling techniques.

7. Others

   Light seismic effects could be envisaged.

8. General assessment

   No major environmental problems seems to be associated with these techniques under normal working conditions if appropriate measures, particularly water recycling, are applied.

GENERAL REFERENCES

H. Baibakov, "How to improve utilization of geological reserves of oil", Oil Engineering, No. 7, July 1974.

Pioneering Ekofisk system to inject gas at 9,200 psi, "Petrol Engineer", 1974, 46/2, 32-34.

M. Grenon, "A propos des ressources mondiales de pétrole: les taux de récupération" Revue Française de l'Energie, 1976, No. 285, 372-377.

USE OF UNDERGROUND NUCLEAR EXPLOSIONS TO INCREASE OIL AND GAS PRODUCTION

Technological aspect

1. Future technology

Method: contained nuclear explosions are triggered under the oil bed at some hundreds of metres below the surface. Measures are taken to prevent any escape of radioactive elements into the atmosphere and to avoid pollution of aquifers. A second well is then drilled in the cavity formed at the place of explosion and oil extracted from the surrounding geological formations. A similar approach is used particularly for tight gas reservoirs.

Economic efficiency: should increase the production of oil wells by 30-50 per cent and gas production up to 6-10 times.

Importance in ECE region: minor. A few pilot experiments.

2. Methods for reducing environmental impact

In principle, the techniques used should avoid any pollution but their failure could have disastrous consequences. In order to decrease the induced radioactivity of the products, oil or gas extracted are delayed from some days to some months; a reduction of the concentration in contaminating long-life isotopes down to safety levels is achieved by dilution with air or clean natural fuel; the fuel extracted with the help of nuclear explosions can also be restricted for utilization in industry. Furthermore no techniques are available to prevent seismic effects, though they can be reduced by using several small charges instead of one powerful one.

3. General assessment

A potentially important technology, particularly for oil extraction. Because of the seismic effects, can only be used in uninhabited regions. The consequences of possible failure of safety measures for preventing radioactive pollution will restrict the potential applications. Underground nuclear explosions can also be used for creating large underground oil and gas storage facilities.

RELATED TOPIC

- Methods of increasing gas or oil extraction efficiency: pp. 38-39.

USE OF UNDERGROUND NUCLEAR EXPLOSIONS TO INCREASE OIL AND GAS PRODUCTION

Environmental aspect

1. Air pollution

The radioactive contamination of ambient air should remain within acceptable levels. No other air pollution except some dust.

2. Water pollution

The radioactive contamination of aquifers should remain within acceptable levels. Some hydrocarbons displaced by the explosion may contaminate water tables.

3. Land use

Practically the same as for traditional methods of oil or gas extraction since the depth at which the explosion occurs should ensure that the ground level effects are negligible.

4. Solid wastes

None

5. Noise

Not significant.

6. Aesthetic aspect

The same as for traditional methods of oil or gas extraction.

7. Others

Possibility of triggering secondary seismic effects. No significant effect on flora and fauna except for some crumbling of rock and banks of brooks and rivers.

8. General assessment

Except for seismic and safety aspects against radioactive pollution no other additional significant environmental aspects than with usual methods for oil or gas extraction.

GENERAL REFERENCES

Toman, John, Status of stimulating tight gas reservoirs with nuclear explosives, "AlGhE Symp. Ser.", 1974, 70, N 142, pp. 171-173.

B. Clenn, K. Field, C. Ford. Process for stimulating petroliferous subterranean formations with contained nuclear explosions, Continental Ore Co., Australia, pat.cl. 80.2.(EOl b), N 432282.

Nuclear explosion extracts oil, Nauka i Zhizn, N 2, 1973.

R.S. Brundage, Environmental Aspects of Nuclear Stimulation of Natural Gas Reservoir Production. "AlGhE Symp. Ser.", 1974, N 142, pp. 176-178.

"IN SITU" EXTRACTION OF COAL (OTHER THAN MINING)

Technological aspect

1.1 Present technology

None currently in use at the industrial level.

1.2 Future technology

Principle: in situ extraction by solvents of the organic components of coal. Generally includes the following steps: hole drilling, injection of an aromatic solvent into the coal seam, dissolution of the organic components, pumping the fluid produced to a processing plant, filtration to remove insoluble materials, solvent recovery and recycling, and desulphurization. The whole process removes all inorganic bound sulphur and 60-70 per cent of the organic sulphur. The principal product is a solid called solvent-refined coal (melting point about 180°C), containing less than 0.1 per cent ash and having a uniform heating capacity of about 37,250 kJ/kg (8,900 kcal/kg), regardless of the type of coal used. It can be used in place of coal with only minor modifications in the combustion equipment. The absence of rock debris is also an important advantage of this method.

Economic efficiency: Low at the present stage of development.

Importance in ECE region: potentially important for the exploitation of currently non-economically recoverable reserves.

2. Methods for reducing environmental impact

Not applicable for the time being.

3. General assessment

This method of coal extraction is attractive but is unlikely to be used in the near future because of its low economic efficiency at the present stage of development.

RELATED TOPICS

- open cast mining of solid fuels: pp. 8-9
- coal fired power plants: pp. 18-21
- coal gasification: pp. 60-61
- coal liquefaction: pp. 62-63

"IN SITU" EXTRACTION OF COAL (OTHER THAN MINING)

Environmental aspect

1. Air pollution

   No detailed information; probably not significant.

2. Water pollution

   No detailed information; probably not significant. Safety measures should avoid potential leakage of solvents into aquifers and local surface water bodies.

3. Land use

   Minor by comparison with traditional methods of coal extraction. Potential risk of land subsidence.

4. Solid wastes

   None to negligible.

5. Noise

   Very limited.

6. Aesthetic aspect

   No serious problems expected.

7. Others

   None.

8. General assessment

   Apparently no significant environmental problem, particularly when compared to traditional methods of coal extraction. This method allows a much cleaner use of coal and therefore appears attractive from the environmental point of view.

GENERAL REFERENCES

A. Hammond, W. Metz, T.H. Maugh II, Energy and the Future, American Association for the Advancement of Science, Washington, D.C., 1973, p.9.

T.H. Maugh II, "Underground Gasification: an Alternative Way to Exploit Coal", Science vol. 198, No. 4322 (1977), 1132-1134.

TAR SANDS UTILIZATION

Technological aspect

1.1 Present technology

Methods of extraction: (i) in situ extraction: steam is injected for about a month into the ground to lower the viscosity of the bitumen deposit. A relatively similar heat effect is attained by a controlled in situ combustion of the bitumen. Pumped oil, gas and water mixture is piped to separators: gas is used as a fuel for steam production, water re-used or returned to a deep formation and oil sent to a refining unit.

(ii) surface mining: extracted sands are treated in an extraction plant where a hot water process separates the bitumen from the sand. The bitumen is subsequently cracked, refined and desulphurized in a processing plant. Sand and other waste materials are returned to the mining site for land rehabilitation.

(iii) underground thermal mining: method operational in at least one project in the USSR which combines the high recovery efficiency of excavation methods with the environmental advantages of in situ extraction processes.

Economic efficiency: lower than traditional methods for oil extraction from oil fields. Recovery factors vary from 12-25 per cent for in situ combustion techniques to more than 50 per cent for mining operations. Total production costs range between $6 and $12/bbl (1976). Requires large investments.

Importance in ECE region: minor at present but tar sands exploitation will certainly play an important role for countries in ECE which have such deposits (particularly Canada, United States and USSR).

1.2 Future technology

Mainly refinements of present technology. Possible use of other solvents for extracting bitumen in situ such as detergents, various alcohols, etc.

2. Methods for reducing environmental impact

Mainly: water recycling measures; usual equipment for air pollution abatement; land rehabilitation (when land restoration is not feasible); and careful planning.

3. General assessment

Great future importance due to the fact that tar sand and oil shale (pp. 46-47) reserves constitute a third of total world recoverable fuel reserves.

RELATED TOPICS

- open-cast mining of solid fuels: pp. 8-9
- methods of increasing gas and oil extraction efficiency: pp. 38-39
- in situ extraction of coal: pp. 42-43
- oil shale utilization: pp. 46-47

TAR SANDS UTILIZATION

Environmental aspect

1. Air pollution

   Emission of large quantities of SO_2, CO_2, NO_x and occasional small emissions of SH_2. These emissions vary greatly with the extraction method used. Dust in case of surface mining.

2. Water pollution

   Mainly consists of dissolved minerals, hydrocarbons and suspended solids. Water pollution problems are minimized by water recycling. A plant with a planned capacity of 125,000 barrels of oil per day would require a maximum of 2.5 m^3/sec. of water.

3. Land use

   Very large, particularly for surface mining operations: a Canadian project with a daily capacity of 125,000 barrels of oil for 25 years will require the mining of more than 200 km^2.

4. Solid waste

   Large amounts of sand mainly used for land rehabilitation when the area has been strip mined.

5. Noise

   Probably considerable (steam injection, excavators, conveyors, etc.)

6. Aesthetic aspect

   Probably significant problem in view of the large areas involved, notably in surface mining operations. Land reclamation is a rule but rehabilitated areas do not necessarily have the same topographic conformation nor the original biological state.

7. Others

   Complete destruction of the indigenous flora and fauna on the surface mined, with limited possibilities for its return after rehabilitation (change in biological conditions). Difficult protection of water tables particularly during *in situ* extraction.

8. General assessment

   The production of oil from tar sands causes definite environmental problems which are mainly a result of the rather gigantic dimensions involved for producing oil at economically acceptable costs.

GENERAL REFERENCES

Department of Energy, Mines and Resources "Oil and Natural Gas Resources of Canada, 1976" and "Oil Sands and Heavy Oils: the Prospects". Reports EP 77-1 and EP 77-2, Printing and Publishing Supply and Services Canada, Ottawa, 1977.

Alberta Oil Sands Environmental Research Programme, various information documents, 9925-107 Street, Edmonton, Alberta, Canada.

Grenon M. "Les ressources non conventionnelles de pétrole et de gaz naturel" *Revue de l'Energie*, January 1976, pp. 22-29.

OIL SHALE[1/] UTILIZATION

Technological aspect

1.1 Present technology

Methods: rocks are heated to 430-600°C; kerogen releases vapours that can be converted to raw shale oil which is further refined into petroleum products.

(i) extraction of oil shale by mining or open-cast mining with primary crushing; transportation, milling, pyrolysis treatment and return of treated rock (about 80 per cent of primary rock) to the place of extraction for its rehabilitation. The final products are synthetic oil and liquefied gas with ammonia, sulphur and carbon residues as by-products.

(ii) *in situ* methods, for example partial excavation of oil shale, blasting of the remaining rock creating a rubble-filled cavern of broken oil shale, igniting the shale at the top of the cavern, support of the combustion process by controlled amount of air pumped in, extraction of oil collected at the bottom of the cavern.

Economic efficiency: *in situ* technology could be economical and seems to range between 10 and 15 US$ per bbl.

Importance in ECE region: potentially important technique for regions where large reserves of shale oil are situated.

1.2 Future technology

Mainly refinements of present technology. Possible use of solvents.

2. Methods for reducing environmental impact

Utilization of gas refinery systems. Refining and recycling of water used for processing.

3. General assessment

Of great importance in the near future due to the large existing oil shale and tar sand reserves (33 per cent of world total fuel recoverable reserves compared to 12 per cent of crude oil reserves).

RELATED TOPICS

- open-cast mining of solid fuels: pp. 8-9
- methods of increasing gas and oil extraction efficiency: pp. 38-39
- *in situ* extraction of coal: pp. 42-43
- tar sands utilization: pp. 44-45

1/ An oil shale is a laminated marlstone rock containing a tar-like organic material called kerogen.

OIL SHALE UTILIZATION

Environmental aspect

1. Air pollution

Total: about 1,000 tons per day for a 1 million barrel-per-day production unit.

NO_x	0.09 kg/bbl
Sulphur	0.5 kg/bbl (as SO_2)
Particulates	0.4 kg/bbl

In addition, usual combustion products should be expected in the case of underground combustion.

2. Water pollution

Very difficult to assess. Water requirements are large (100-150 million m^3/year for a 1 million barrel-per-day operation) and may require watercourses diversion.

Chemical pollution of water bodies due to erosion and leaching of spent shale deposits. Increased salinity of rejected water brines. Accidental spills of solvents and oil.

3. Land use

Considerable. Cumulative land requirement over 30 years may reach more than 250 km^2 for a 100,000 bbl/day extraction capacity.

4. Solid wastes

Very large quantities. Can be used for backfill and land restoration or deposited in abandoned underground sites.

5. Noise

High noise level due to blasting, earth-moving equipment, crushing and grinding operations, compressors, pumps, processing operations, etc. involved in this massive industrial operation.

6. Aesthetic aspect

Significant: destruction of flora on the site and strong impact on fauna. If appropriate tourism might be expected to decline. Possible land subsidence.

7. Others

Lowering of water tables in the vicinity of the mine if underground water is used.

8. General assessment

Important environmental problems requiring careful planning and operation, particularly in view of the considerable land use and aesthetic aspects.

GENERAL REFERENCES

"Environmental considerations in future energy growth", Battelle, for the United States Environmental Protection Agency, Vol. 1, (1973).

H. William, "Processed shale disposal for a commercial oil shale operation", Mining Congr. 1974, 60, N 5, pp. 25-29.

L. Grainger, "The Robins Coal Science Lecture 1974: Coal into the Twenty-first Century", Journal of the Institute of Fuel, 66 June 1975.

GAS TURBINES

Technological aspect

1. Present technology

Method of production: the gas-turbine cycle is based on the almost adiabatic compression of air, the almost constant-pressure combustion of fuel with that air, and the adiabatic expansion of the hot gases back to atmospheric pressure. In the simplest arrangement the compressor and turbine are on the same shaft, and the difference between the work done by the turbine and that required by the compressor is the net work output. The thermal efficiency attained seems to range between 18 per cent ("simple cycle") to about 28 per cent (regenerative cycles). Also often used as compressors along pipelines and in aircraft engines. Gas turbines must use low sulphur fuels for technical reasons.

Economic efficiency: the relatively low efficiency of gas turbines for electricity generation made their use for extended periods quite uneconomic; however, their very fast start-up times (compared with those for steam units) and low investment costs were ideal for peaking units. More recently, the quick delivery time of gas turbines in the face of construction and their siting flexibility (as they do not require cooling water) are moving them into the base-load category. Advanced design studies of combined gas turbine/steam plants show over-all thermal efficiencies of about 50 per cent against 40 per cent for modern steam units.

Importance in ECE region: no data available on importance in electric power plants. A significant proportion of the gas pipelines compressors are driven by gas turbines.

2. Methods for reducing environmental impact

(i) optimization of combustion in order to minimize air pollution. NO_x emissions are mainly resulting from the high combustion temperature;

(ii) as a noise abatement measure both air inlet and exhaust gas outlets must be carefully arranged and provided with efficient silencers. Residential buildings should be located at least 700 m away from the compressor and turbine station.

3. General assessment

Present use of gas turbines for stationary power plants is confined largely to peaking power applications. Development in gas turbines for aeronautical purposes, however, makes it increasingly appropriate to consider 200-500 MW base-load gas turbines for 1,000 MW stations and advanced open-cycle systems using both gas and steam turbines. The efficiency increase (over those for modern steam units) may be high enough to justify the extensive fuel treatments required to adapt most fuels for gas turbine use.

RELATED TOPICS

- gas pipelines: pp. 12-13
- gas-fired power plants and other fossil-fuelled power plants: pp. 14-21
- coal gasification: pp. 60-61
- biomass energy: pp. 92-93

GAS TURBINES

Environmental aspect

1. Air pollution

Approximate amounts (tons) released annually per 1,000 MWe

CO	1,100 - 2,700	SO_x	11,300
Hydrocarbons	450	Particulates	Variable
NO_x	11,300-72,500	Radioactive	None

2. Water pollution

None. Recent stations are air-cooled avoiding waste heat discharges to watercourses.

3. Land use

No large site requirements because of their compact size (in some cases the units are even regarded as mobile and are moved from site to site as requirements demand).

4. Solid wastes

None.

5. Noise

Can be a serious problem if silencing equipment is not installed.

6. Aesthetic aspect

Most gas turbines have no visible exhaust plume.

7. Others

None.

8. General assessment

Pollutants emitted from turbine engines appear to differ substantially from conventional boiler emissions in that NO_x, CO and hydrocarbon emissions may be considerably higher per unit of power generated.

Noise pollution may become an important problem if considerably larger generating units were to be built in urban areas.

GENERAL REFERENCES

H.C. Hottel and J.B. Horvard, New Energy Technology. Some facts and assessment, The Massachusetts Institute of Technology Press, Cambridge, Mass., 1971.

"Engineering for Resolution of Energy-Environment Dilemma", National Academy of Engineering, Committee on Power Plant Siting, Washington, D.C., 1972, p. 74.

NEW HEAT TRANSFER MEDIA

Technological aspect

1.1 Present technology

Purposes: (a) to increase the conversion efficiency of thermal power plants by transferring heat from the boiler at higher temperatures than is technically feasible with water;
(b) to allow the use of small temperature differences for producing electricity.

Techniques: in advanced power plants such as breeder reactors (pp. 58-59) or HTGR (pp. 30-31) molten sodium, helium or carbon dioxide transfer heat from the reactor core to heat exchangers where steam is produced. Ammonia and chlorofluoromethanes (freons) are the most commonly indicated heat transfer media for converting into electricity some renewable sources of energy such as ocean temperature gradients (pp. 98-99); the heat transfer fluid directly activates turbines.

Economic efficiency: demonstrated for CO_2; not fully demonstrated but highly probable for helium and sodium. To be demonstrated for others.

Importance in ECE region: limited so far to GCR and HTGR (pp. 28-31) and to several types of industrial demonstration plants.

1.2 Future technology

Organic compounds particularly for low temperature application.

2. Methods for reducing environmental impact

Increased safety measures. Improved containment techniques. Development of heat transfer media which presents no environmental hazard.

3. General assessment

Technology in full development which will very likely play an important role in the future.

RELATED TOPICS

- thermal power plants (fossil- and nuclear): pp. 14-31 and 58-59
- solar steam power plants: pp. 82-83
- utilization of ocean temperature gradients: pp. 98-99

NEW HEAT TRANSFER MEDIA

Environmental aspect

1. Air pollution

   None to very limited during normal operations.

2. Water pollution

   None to very limited during normal operations.

3. Land use

   Irrelevant.

4. Solid waste

   Irrelevant.

5. Noise

   No particular problem.

6. Aesthetic aspect

   Irrelevant.

7. Others

   Hazards in case of accidental releases: sodium explodes when it comes into contact with water and ignites with air; ammonia is toxic and chlorofluoromethanes are believed to endanger the ozone shield. Helium and carbon dioxide releases would present no environmental hazard.

8. General assessment

   Field of technology which could significantly reduce the amounts of waste heat released to the environment and allow the tapping of some renewable sources of energy. The development of harmless heat transfer media should be encouraged.

MAGNETOHYDRODYNAMIC GENERATION

Technological aspect

1.1 Present technology

Method of production: MHD generation of power is based on the same principle as conventional generation. Instead of a solid conductor (turbine rotor) moving across a magnetic field, a jet of ionized fluid is forced through it. By placing electrodes in the fluid stream, direct current electricity at relatively high potential, e.g. 2,000 volts or more, can be obtained. The substitution of this ionized fluid for the armature of a conventional turbo-generator characterizes MHD generation. In the open-cycle system, the working fluid consists of the gaseous products of fossil fuel combustion, seeded with an easily ionized element, such as potassium or cesium. A number of different types of MHD cycles have been considered, but the consensus is that the most likely near-term hope for using MHD to produce central station power is the open-cycle fossil-fuelled system.

Thermal efficiency: Over-all plant efficiency of 50-60 per cent is predicted when MHD is used as a topping cycle with a conventional steam generator for a.c. power generation.

Economic efficiency: Reliable estimates of the cost of MHD plants are not yet available, but there appears to be general agreement that construction costs should be about the same as for traditional coal-fired plants. Operating costs per kW could be lower because of the more efficient use of fuel.

Importance in ECE region: Research on MHD is becoming widespread, with active efforts in the USSR, United States and several European countries. The Soviet Union currently has the world's largest open-cycle MHD power station (at the pilot stage); it operates on natural gas and produces 25 MW. A 1,000 MW unit is to be built by 1981.

1.2 Future technology

Closed-cycle plants using liquid metals or gas in connexion with nuclear reactors are in the research stage.

2. Methods for reducing environmental impact

No specific measure seems applicable for NO emissions. SO_x emissions can be controlled by seeding flue gases with potassium in sulphur-free form and collecting K_2SO_4 with filters or electro-precipitators.

3. General assessment

MHD is a completely new technology: no moving parts in generator; no steam cycle; rapid start-up time; and can build extremely large plants.

Numerous problems must be overcome before MHD can become a viable alternative to central station generation, including economic recovery of the ionization seed, gas conductivity, materials, control of air pollutants. Channel insulation, electrode fabrication, and superconductive magnets are also formidable engineering challenges. Significant progress has recently been achieved in this field.

RELATED TOPICS

- fossil-fuelled power plants: pp. 14-21
- nuclear power plants: pp. 22-31 and 58-59
- new heat transfer media: pp. 50-51

MAGNETOHYDRODYNAMIC GENERATION

Environmental aspect

1. Air pollution

The NO concentration in the stack gas from MHD power plants is expected to be very high, because of the unusually high temperature of the combustion gases. SO_x emissions are significantly high but practically all the sulphur can be removed, even from high-sulphur coals.

2. Water pollution

Reduced waste heat discharges - when compared with usual conversion techniques - due to higher theoretical conversion efficiency.

3. Land use

Considered as a minor problem.

4. Solid waste

Various negative effects of coal slag (if coal is used as a fuel).

5. Noise

Could be a significant problem if gas turbines (pp. 48-49) were used as a second half of an MHD power plant.

6. Aesthetic aspect

No particular problem. Structure contained in power plant buildings.

7. Others

No specific information available.

8. General assessment

The main environmental problems appear to be nitric oxide formation in high concentrations and various effects of coal slag.

Present data on the over-all environmental impact of MHD generation, when compared to that of a conventional plant are inconclusive. Small-scale research on gas cleaning and seed recovery should focus on the determination of NO, NO_2, SO_2 and SO_3 to be expected at MHD steam plant exit, the measurement of chemical and physical properties of slag-seed mixture and deposition rates, and studies of methods for removal of NO_x and SO_x from flue-gas.

MHD is potentially a valuable method for burning coal with a high sulphur content.

GENERAL REFERENCES

"Vestnik of the Academy of Science of the USSR", Vol. 2, February 1975, p. 12.

"Engineering for resolution of the energy-environment dilemma", National Academy of Engineering, Committee on Power Plant Siting, Washington, D.C., 1972.

A. Hammond, W. Metz and T. Maugh II, Energy and the Future, American Association for the Advancement of Science, Washington, D.C. (1973), p. 27.

FUEL CELLS

Technological aspect

1. Present technology

Method of production: in fuel cells, the chemical energy produced by the oxidation of a gaseous fuel is converted directly into electricity. In theory, the efficiency of this conversion has no fundamental limitations.

The following gases may be used as fuels: H_2, N_2H_4, CH_3CH, and natural gas; O_2, H_2O and air are used as oxidants. Combustion takes place in an electrolyte which may be a ceramic solution or an aqueous solution of crystals of hydrates (e.g. Baur electrolyte ($KOH.H_2O$)). Combustion occurs at temperatures between 130°C and 1,000°C; its by-products are CO_2, H_2O and NO_x.

Thermal efficiency: about twice that of conventional generators.

Economic efficiency: This technology is not yet competitive with conventional technologies, but if offers very extensive possibilities for the production of electric power, particularly from hydrogen.

Importance in ECE region: experimental plants in France, Germany, Federal Republic of; United States and USSR. Practical applications at present limited to space exploration. Of very great importance for the future, since fuel cells open the way to an efficient use of hydrogen as a fuel for the production of electric power (pp. 64-65).

2. Methods for reducing environmental impact

None required.

3. General assessment

Fuel cells avoid the complications which arise from boilers, turbines and generators in thermal power plants; and, in their case, the Carnot cycle is not involved. There is no theoretical limit to their efficiency. Small mobile power sources are likely to be developed in the near future; the development of centralized plant with a higher capacity is possible.

To bring this technology to optimum utilization in electric power systems, research will have to continue on a number of aspects, such as the life of the cells, the conversion of the fuel and the choice of electrolytes.

RELATED TOPIC

- hydrogen production and use: pp. 64-65.

FUEL CELLS

Environmental aspect

1. Air pollution

Virtually no problems arise. Maximum emissions (in kg of pollutants per 1,000 kWh) for an experimental fuel cell.

SO_2	0.000121	Hydrocarbons	0.105
NO_x	0.110	Particulates	0.000014

2. Water pollution

None.

3. Land use

Compared to other methods of electric power production, fuel cells do not require large areas. Used as decentralized power sources, the requirements for transmission would be reduced, with consequent savings in land use.

4. Solid waste

Nil to insignificant.

5. Noise

The noise is localized in the industrial premises where the compressors, etc., are installed.

6. Aesthetic aspect

Deposits of fuel (tanks).

7. Other problems

Installations for the transport and storage of gases may explode. The safety measures to be taken to counter this risk are known, and the problem is a minor one.

8. General assessment

In principle, fuel cells may be regarded as a technology which is very favourable from the environmental point of view.

GENERAL REFERENCE

A. Hammond, W. Metz and T. Maugh II: Energy and the Future, American Association for the Advancement of Science, Washington D.C. (1973), p. 160.

FLUIDIZED BED COMBUSTION

Technological aspect

1.1 Present technology

Method: combustion at relatively low temperatures (500°C-750°C) of small particles of solid or liquid fuel in a state of suspension and forming a "bed" supported by jets of air emitted under pressure from several orifices and directed from below upwards. A pipe manifold, with water and steam circulating within it, is immersed in this "bed" and transfers the heat to its place of use (for example, the turbine). Technological advantages: small size, efficient temperature control, little corrosion, possibility of coupling to gas turbines by using equipment operating under pressure.

Thermal efficiency: about 70 per cent of the heat produced is transferred to the water circulating in the pipe manifold.

Economic efficiency: considerable saving in investment costs (estimated at 21 per cent for a 660 MWe thermal electric power plant) and in operating costs (10-14 per cent without desulphurization and 12-16 per cent with desulphurization).

Importance in ECE region: experimental and semi-industrial installations in several countries. Industrial use now beginning. Great potential importance, since fluidized bed combustion makes it possible to use - without prior desulphurization - coal with high sulphur content (more than 1 per cent; coal in this category accounts, for example, for 80 per cent of United Kingdom coal and 65 per cent of United States coal).

1.2 Future technology

Adaptation of the present technology to high capacity plant.

2. Methods for reducing environmental impact

The addition of limestone and dolomite in the fluidized bed makes it possible to elminate up to 90 per cent of the sulphur in the fuel. The sulphur is collected in the form of sulphates of Ca or Mg with the ash. The introduction of this method accounts for less than 1 per cent of the capital invested and about 3 per cent of production costs.

3. General assessment

With fluidized bed combustion, it is possible to use coal with a high sulphur content, without preliminary treatment and with an investment cost which is lower than that for conventional power plants. It would appear that this technology is ready for industrial use and is destined for extensive development.

RELATED TOPICS

- fossil-fuelled power plants: pp. 14-21
- gas turbines: pp. 48-49
- new heat transfer media: pp. 50-51
- coal gasification and liquefaction: pp. 60-63

FLUIDIZED BED COMBUSTION

Environmental aspect

1. Air pollution

CO: limited amounts since combustion takes place in air saturation conditions.

Hydrocarbons: probably no very considerable emissions.

NO_x: not very considerable, owing to low temperature of combustion: depends on the fuel: 60 to 180 ppm for coal and 80 to 120 for fuel oil.

Sulphur: varies according to sulphur content of the fuel. 90 per cent of the sulphur can be eliminated by adding limestone or dolomite to the bed.

Particulates: small quantities (temperature maintained below the point of fusion of the ash).

Radioactivity: nil to negligible (coal).

2. Water pollution

None. For conversion into electric power, see pp. 14-21.

3. Land use

No problem. The plant is smaller, for the same capacity, than conventional boilers.

4. Solid waste

Varies, depending on the fuel. The same order of magnitude as for other fossil fuelled power plants (pp. 14-21).

5. Noise

No specific information.

6. Aesthetic aspects

No problem, since the plant is located inside the power plant.

7. Others

None.

8. General assessment

Compared with other types of fossil fuel boilers, the fluidized bed combustion technique offers some definite advantages in regard to the environment. Its use on a large scale would be a major element in the preservation of the quality of the environment.

GENERAL REFERENCES

H.B. Locke, Fluidized combustion for advanced power generation with minimum atmospheric pollution, Journal of the Institute of Fuel (190), September 1974.

Fluidized bed combustion of coal (report submitted by the United Kingdom, ECE Coal Committee, COAL/WP.4/Working Paper No. 17, 30 March 1971).

L. Grainger, Zukünftige Verfahren der Kohlenumwandlung, Glüchauf 111 (1975), No. 14.

H.R. Hoy and H.B. Locke, Research and Development Work on Fluidized Bed Combustion, Swedish Institute of Engineers, October 1973, Stockholm, Sweden.

FAST BREEDER REACTORS

Technological aspect

1.1 Present technology

Method of electrical generation: heat is produced by a controlled fission chain reaction in a fuel composed of a mixture of plutonium and uranium oxides clad in stainless steel and packed in a highly compact core. FBR are distinctive from other nuclear reactors by the absence of a moderator resulting in a fast chain reaction providing an intense neutron flux from the edge of the core. These high energy neutrons (fast neutrons) are captured by a core-surrounding blanket of "depleted" uranium-238 which is transmuted to plutonium, making this type of reactor produce its own fuel. The most advanced FBR are cooled by molten sodium, a very efficient metal as a heat transfer medium kept at atmospheric pressure at about 550°C (for use of sodium as coolant, see pp. 50-51). A primary heat exchanger heats a secondary liquid sodium circuit containing steam generators. Steam then drives turbines or is used for other industrial purposes.

Conversion efficiency: the FBRs in the ECE region have an average 37 per cent thermal conversion efficiency (1976).

Economic efficiency: not fully demonstrated as yet as all operational FBRs are still prototypes. FBRs make use of cheap U-238 stocks but are highly dependent on reprocessing facilities. Very large investments are required.

Importance in ECE region: in 1976 the installed capacity of ECE in FBR was 2,358 MWth and 876 MWe.

1.2 Future technology

Technical improvements for handling hot sodium under various conditions. Potential development of other heat transfer media. Safety improvement notably for fast and reliable heat removal from a very compact core.

2. Methods for reducing environmental impact

Very limited information available. Apparently quite similar to those indicated for PWR (pp. 22-23) with particular emphasis on safety aspects, notably during fuel transport. Selection of construction materials for minimizing the amounts of radioactive materials produced by neutron activation.

3. General assessment

FBRs probably represent one of the most sophisticated industrial technologies ever developed and, as all FBRs are prototypes, investment costs are huge for the time being. However if a "breeder economy" is chosen, they should considerably decrease. FBR have a low dependence on the uranium ore market but highly rely on an appropriate reprocessing capacity. The future of this type of reactor in the ECE region will mainly result from political decisions and from the availability and efficiency of reprocessing facilities.

RELATED TOPICS

- other types of nuclear reactors: pp. 22-31
- new heat transfer media: pp. 50-51

FAST BREEDER REACTORS

Environmental aspect

1. Air pollution

   Probably similar to, or comparable with, PWR (pp. 22-23).

2. Water pollution

   Same as for PWR.

3. Land use

   Comparable to, or smaller than, PWR.

4. Solid waste

   Comparable to, or smaller than, PWR.

5. Noise

   No information available; probably similar to PWR.

6. Aesthetic aspect

   Comparable to PWR.

7. Others

   Same features as for PWR (pp. 22-23). Expressed safety concerns are generally stronger than for other types of nuclear reactors. Hot sodium handling (pp. 50-51). Decommissioning.

8. General assessment

   Comments made about PWR (pp. 22-23) are probably also valid for FBR. Strong concerns about the environmental implications of abnormal operating conditions of FBR have been expressed in several ECE countries.

COAL GASIFICATION

Technological aspect

1.1 Present technology

Method: the carbon of coal is combined with steam at high temperature and pressure to form methane (general formula is: coal + steam →CH_4 + CO_2); in general the gasifier operates at 20-70 bars and up to 1,500°C and produces a synthetic gas containing 40-65 per cent methane. Contaminants (CO_2, sulphur compounds and water vapour) are then removed from the gas which is subjected to a catalytic methanization. The purified final product is called synthetic natural gas (SNG). Differences between various processes include mainly the manner in which coal is admitted to the gasifier, the type of reactor bed (fixed or fluidized), the heat source for gasification. Processes also vary according to the type of coal used (lignite, sub-bituminous, etc.). SNG does not require special gas burners.

Economic efficiency: costs vary considerably with many factors. SNG from coal-Lurgi process (Federal Republic of Germany) : 4.1 to 4.5 US$ per million kcal 1/; SNG from coal - US process (United States) : 2.7 to 3.8 US$ per million kcal. Prices in the range of 15.0 US$ per million kcal have been recently estimated. Plant cost for a daily production of 6.8×10^{10} kcal : about 1 billion US$.

Importance for ECE region: the production of SNG from coal or other fossil fuels becomes of growing importance in the ECE region in view of an increasing demand for gas, low natural gas reserves and the quite large deposits of low-grade coal which are available in the ECE region. Capacity to be installed in 1985 in the United States only : 12 to 37 plants with a total daily production of more than 250 million m^3.

2. Methods for reducing environmental impact

Utilization of ashes and sulphur in other economic sectors.

3. General assessment

Because of its importance in the ECE region, the use on a large scale of new technologies for SNG production is expected to be accelerated. This is in spite of the low economic efficiency of SNG production in comparison to traditional methods of the extraction of natural gas. The price for SNG and NG will level in the future because of the further reduction of NG reserves and further improvement in the technologies of SNG production.

RELATED TOPICS

- open-cast mining of solid fuels: pp. 8-9
- gas pipelines: pp. 12-13
- gas fired power plants: pp. 16-17
- in situ extraction of underground coal: pp. 42-43
- gas turbines: pp. 48-49
- fluidized bed combustion: pp. 56-57
- coal liquefaction: pp. 62-63

1/ 1 kcal = 4.185 kJ

COAL GASIFICATION

Environmental aspect

1. Air pollution

   Amounts and nature vary with the technique used. The main emission is SO_2.

2. Water pollution

   No specific information available; in most coal gasification processes, waste waters are treated and recycled.

3. Land use

   It is estimated that a 250 x 10^9 kJ/day (output) gasification plant able to supply gas to a 1,000 MW power station would require between 0.7 and 0.8 km^2.

4. Solid wastes

   Large quantities composed mostly of ashes (vary with ash content of coal) and from compounds used to remove sulphur. If not sold, the elemental sulphur produced should be considered as a solid waste.

5. Noise

   No particular problem due to the plant itself but transport devices may create difficulties (cranes, railroads, conveyor belts, etc.).

6. Aesthetic aspect

   No particular problem.

7. Others

   None specific to coal gasification facilities.

8. General assessment

   Ashes seem to be the major environmental problem; the possibilities of using them in other economic sectors is under examination. Rehabilitation of mined areas (pp. 8-9).

GENERAL REFERENCES

"Environmental considerations in future energy growth", Battelle, for the United States Environmental Protection Agency, Vol. 1, (1973).

A.L. Hammond, W.D. Metz and T.H. Maugh II, "Energy and the Future", American Association for the Advancement of Science, Washington, D.C., 1973, p. 12.

O. Hammon and M.B. Zimmerman, "The economics of coal-based synthetic gas", Technology Review. Massachusetts Institute of Technology, Cambridge, Mass., July/August 1975, pp. 42-51.

Economic Commission for Europe, Coal Committee, Symposium on Coal Gasification and Liquefaction, Dusseldorf, Federal Republic of Germany, 12-16 January 1976; report: COAL/SEM.3/2; list of documents COAL/SEM.3/R.1. A second symposium to be held in Katowice, Poland, April 1979 (see COAL/SEM.6/Inf.1 and 2).

COAL LIQUEFACTION

Technological aspect

1. Present technology

Methods of production:

(i) multistage pyrolysis process in which coal is heated in the absence of oxygen to produce a mixture of oil, volatile components, charcoal residue and a gas having a low heating capacity. After oil separation and a treatment with water, a new product called synthetic crude oil is obtained. The volatile components which are removed in the process are used as feedstocks in chemical processes.

(ii) coal is treated as described for the preparation of "solvent refined coal" (see pp. 42-43) except that the organic material produced is sufficiently hydrogenated during the extraction to remain in a liquid form. Part of the extracted material is recycled as fresh solvent. The remainder is treated with a catalytic hydrocracking process to produce a synthetic crude oil.

Note: various liquid coal mixtures could also be mentioned here although formally their preparation cannot be considered as a process of coal liquification. For example: coal slurry (ratio of coal and water 1 : 1) or a mixture of pulverized coal, oil and water (optimum ratio: 6 : 2 : 2), allowing boilers designed for burning oil to be converted to coal with little modification. These technologies have a number of economic and environmental advantages if compared with traditional methods of coal transportation and burning and they are expected to be widely used in the ECE region in the future. (See also pp. 68-69).

Economic efficiency: The projected cost for oil and charcoal produced by pyrolysis and oil produced by solvents is apparently too high to justify the use of these technologies even in large demonstration plants.

Importance for ECE region: Minor. Only a few experimental installations now in operation. These methods are potentially important if new technologies to reduce costs would be developed.

2. Methods for reducing environmental impact

Generally included in the present technology. No additional measure seems presently to be required.

3. General assessment

The industrial preparation of synthetic crude oil will become commercially viable only when the costs of its preparation become considerably lower or if the price of natural crude oil were to rise dramatically. Neither prospect seems likely within this decade.

RELATED TOPICS

- open cast mining of solid fuels: pp. 8-9
- *in situ* extraction of underground coal: pp. 42-43
- tar sands and oil shales exploitation: pp. 44-47
- slurry pipelines: pp. 68-69

COAL LIQUEFACTION

Environmental aspect

1. Air pollution

Mainly NO_x, CO, particulates and hydrocarbons resulting mostly from the combustion of coal as fuel. About 99.9 per cent of the sulphur could be recovered.

2. Water pollution

If water is recycled, water emissions from the plant should be minimal.

3. Land use

Estimated at 3.2 km^2 for producing the amount of synthetic crude oil required to generate 1,000 MW.

4. Solid wastes

Mostly process slag; varies with the ash content of the coal used.

5. Noise

Should not be an important problem.

6. Aesthetic aspect

No particular problem.

7. Others

None.

8. General assessment

This technology seems attractive from an environmental viewpoint as the pollution problems are limited particularly in view of the high sulphur recovery efficiency.

GENERAL REFERENCES

"Environmental considerations in future energy growth", Battelle, for the United States Environmental Protection Agency, Vol. 1 (1973).

L. Grainger, "The Robins Coal Science Lecture 1974: Coal into the Twenty-first Century", Journal of Institute of Fuel, 66, June 1975, p. 70.

A. Hammond, W. Metz and T. Maugh II, Energy and the Future, American Association for the Advancement of Science, Washington, D.C., 1973, p. 9.

"Liquid Coal could Replace Oil", Electric Review, 22 November 1974, p. 660.

Economic Commission for Europe, Coal Committee, same reference as for coal gasification (p. 61).

HYDROGEN PRODUCTION AND USE

Technological aspect

1.1 Present technology

Methods of production:

(a) Electrolysis of water: water is dissociated into H_2 and O_2 by DC. Easy recovery of H_2 at the anode. Technology applicable to all large-scale sources of electricity. Potentially applicable to the storage of energy produced by intermittent source of energy (particularly renewable ones).
Theoretical efficiency: 2.79 kWh per m^3 of H_2 in gaseous form.
Practical efficiency: 60 per cent.

(b) Steam reforming with hydrocarbons: general formula:
$CH_2 + 2H_2O \rightarrow CO_2 + 3H_2$. H_2 must be purified.

Cryogenic storage in liquid form at 20°K or below (losses approximately proportional to $[\text{volume}]^{-1/3}$) and transported as liquid in dewars or compressed gas in steel cyclinders.

Economic efficiency: when originated from conventional sources of energy, will always be more expensive than these although its use may be advantageous for some specific applications (urban distribution, etc.). Presently used for space experiments.

1.2 Future technology

Method of production: thermal dissociation of water at about 2,500°C. Potential application for fusion power (pp. 90-91). Further purification of H_2 is required.

Method of storage: in metal hydrides "batteries" (as for example: $Hg_2Ni + 2H_2 \rightleftharpoons Mg_2NiH_4$); synthetic organic or inorganic hydrides might provide better acceptance capacity.

Method of transportation: piprlines similar to gas pipelines (pp. 12-13). As heating value of H_2 (12,100 kJ (2,892 kcal)/m^3 of gas) is three times lower than natural gas, three times more hydrogen should be transported to provide the same amount of energy. Cost of transportation: between 0.033\$/10^6BTU per 100 miles (long distance pipes) to 0.066\$/10^6BTU for distribution to consumers. 1/

2. Methods for reducing environmental impact

None are specifically required.

3. General assessment

Hydrogen might be widely used in the future for industrial and domestic applications. It can be generated from both conventional and renewable sources of energy. Its transportation through piepline does not cause any serious problem (in fact the widely used coal gas contains nearly 50 per cent hydrogen). However its applications for automotive transportation means (such as car, aircraft, cargo, etc.) is limited by safety measures (explosions, etc.) or by the development of metal hydride batteries.

RELATED TOPICS

- gas pipelines: pp. 12-13
- fuel cells: pp. 54-55
- photovoltaic conversion: pp. 86-87

1/ Corresponding respectively to 0.08 and 0.16\$/10^6 kcal (=4.185 kJ) per 100 km.

HYDROGEN PRODUCTION AND USE

Environmental aspect

1. Air pollution

No problem when $H_2 + 2\ O_2 \rightarrow 2H_2O$ (fuel cells, see pp. 54-55). Some NO produced when H_2 is burned using atmospheric oxygen. Heat pollution when burnt as a gas.

2. Water pollution

None.

3. Land use

Varies with the primary source of energy (see respective chapters).

4. Solid wastes

Not significant. Used hydride batteries should be recycled as they contain potentially polluting chemicals (mercury).

5. Noise

Localized: liquefaction installations, compressors, etc.

6. Aesthetic aspects

Storage of liquid H_2 is usually in spherical white tanks. Not important.

7. Others

(a) explosion risk in gaseous and liquid form even with small but intense source of heat (cigarette, for example) as auto-ignition temperature is 585°C. Problem not encountered with hydrides "batteries".

(b) with metal hydrides "batteries" handling of significant amounts of H_2, would require careful manipulations.

8. General assessment

No significant environmental problem. No depletion of natural resources if water is used as chemical source of hydrogen.

GENERAL REFERENCES

W. Winsche and coll., "Hydrogen and its Future Role in the Nation's Energy Economy", Science, Vol. 180, No. 1325 (1973).

L. Jones, "Liquid Hydrogen as a Fuel for the Future" Science, Vol. 174, No. 367 (1971).

R. Wentorf and R. Hanneman, "Thermochemical Hydrogen Generation", Science, Vol. 185, No. 311 (1974).

BATTERY STORAGE OF ELECTRICITY

Technological aspect

1.1 Present technology

Principle: Reversible conversion of electrical energy into chemical energy. Existing batteries operate at ambient temperature. All of them produce continuous current.

Principal types: (a) lead-acid: main application: starting combustion engines; efficiency: 80 per cent; low theoretical specific energy (0.2-0.4 kWh/kg); do not withstand full cycles (charge-discharge) very well; annual cost: about $75 per kW.

(b) nickel-iron; main application: electric traction; efficiency: 60 per cent; average theoretical specific energy; withstand full cycles; annual cost; $95 per kW.

(c) nickel-cadmium; main application: electronics; efficiency: 65 per cent; high theoretical specific energy; withstand full cycles; annual cost $150 per kW.

Economic efficiency: Suitable for specific uses because batteries are movable but they cannot at the present time store large quantities of electricity economically nor are they competitive with, for example, storage lakes (pp. 32-33) and gas turbines (pp. 48-49).

Importance in ECE region: None as regards storage of large quantities of electricity; some experimental installations.

1.2 Experimental and future technologies

Principle: Exactly the same as for existing batteries. Batteries often operating at high temperature. Main potential applications: traction and industrial storage.

Principal types: (a) sodium-sulphur: contain fused sodium and sulphur electrodes; the electrolyte is a strong ceramic, β-alumina ($Na_2O.11Al_2O_3$); efficiency: 60 per cent; theoretical specific energy: 0.75 kWh/kg; withstand full cycles fairly well; annual cost: about $50 per kW; operate between 300 and 350°C.

(b) lithium-sulphur: contain an eutectic mixture of fused lithium salts; efficiency: over 80 per cent; theoretical specific energy: 2.6 kWh/kg; withstand full cycles fairly well; annual cost: $25-50 per kW; operate at 400°C.

Experimental work underway on other types of batteries (lithium-chlorine, zinc-chlorine, zinc-potassium, titanium-lithium, etc.).

2. Methods for reducing environmental impact

None required.

3. General assessment

Wide potential use once the experimental batteries are fully developed and their cost is brought down to $20 per kWh: transport, storage to meet peak demand, storage of renewable sources of energy (solar energy, wind power, etc.) and so forth.

RELATED TOPIC

- hydrogen production and use: pp. 64-65.

BATTERY STORAGE OF ELECTRICITY

Environmental aspect

1. Air pollution

   Lead-sulphuric acid batteries give off a little SO_2 and hydrogen (danger of explosion in enclosed space).

2. Water pollution

   None.

3. Land use

   Low in comparison with storage lakes, for example. A 100 MWh installation would require about 8 m^3 of batteries.

4. Solid waste

   None. In principle, the components of these batteries can be recycled after use.

5. Noise

   None.

6. Aesthetic aspect

   No serious problem.

7. Others

   Industrial manufacture of the more common existing types of batteries entails the use of large quantities of lead and cadmium; so that great care must be taken to protect workers and the environment from these heavy metals.

8. General assessment

   Storage batteries, particularly those now being developed, offer undeniable advantages and their use raises practically no environmental problems. On the other hand, the manufacture and the disposal of existing types of batteries call for constant precautions because of the toxic nature of some of their components.

GENERAL REFERENCES

A.L. Robinson, "Advanced storage batteries: progress, but not electrifying", Science, Vol. 192, 1976, pp. 541-543.

A.L. Robinson "Energy storage (I): using electricity more efficiently", Ibid., Vol. 184, 1974, pp. 785-787.

G.C. Gardner, and others "Storing electrical energy on a large scale", C.E.B.G. Research, May 1975, No. 2, pp. 12-20.

M.S. Whittingham, "Electrical energy storage and intercalation chemistry", Science, Vol. 192, 1976, pp. 1126-1127.

SLURRY PIPELINES

Technological aspect

1. Present technology

The solid to be transported is ground to a predetermined and closely controlled size and mixed with a carrier liquid such as water or oil. Thereafter the slurry is handled much like a pure liquid. Several technical problems are encountered:

(a) the permissible slope of the pipeline is reduced from that of normal liquid pipelines because of a tendency for the solid to settle at the bottom of the slope and block the pipe;

(b) when the system is shut down the solids settle and may cake;

(c) abrasion of the pipeline and pumps makes maintenance costs high;

(d) grinding action of the solid produces "fines" which are difficult to remove from the carrier liquid;

(e) chemicals additives used to keep the solid in suspension are difficult to remove or deal with at the receiving end.

Economic efficiency: Slurry pipelines are effective when there is a large long-term demand for a solid, for instance, 10 million tons of coal for 10 years.

Importance in ECE region: Slurry pipelines are not used extensively anywhere but are being used successfully for the transportation of coal and various minerals (limestone, copper concentrate, magnetite and hematite, etc.).

2. Methods for reducing environmental impact

Use of biodegradable suspension agents for water carrier and organic suspension agents for oil carrier. Recycling of the carrier. Back-up safety systems for avoiding accidental slurry releases.

3. General assessment

Slurry pipelines have a good potential where a large, stable demand exists. Some further developments and improvements would be required particularly in the preparation and recovery of the slurry, the design of pumps, the removal of fines and the recycling of carrier liquids.

RELATED TOPICS

- gas pipelines: pp. 12-13
- capsule pipelines: pp, 70-71

SLURRY PIPELINES

Environmental aspect

1. Air pollution

Potential emissions of particulates during grinding operations of the solid to be transported. Heat releases if drying of solid is required.

2. Water pollution

Chemical pollution depending on type and quantity of chemical used to maintain the suspension. Particulates associated with water carrier: large settling ponds would be required to remove fines. Flocculants might be necessary. Possible erosion during construction of the pipeline. Accidental slurry releases.

3. Land use

As for other pipelines (see pp. 12-13).

4. Solid wastes

None to minor.

5. Noise

Not serious, except at compressors or vacuum pump stations and during construction.

6. Aesthetic aspect

No particular problem if the pipeline is buried and the ground revegetated.

7. Others

In some areas the requirement for large quantities of water to transport solids may be detrimental to local water supplies. Migrations of wild fauna may be affected if pipelines are above the ground.

8. General assessment

The use of slurry pipelines will have acceptable environmental effects when the solid recovery and carrier liquid treatment problems are adequately dealt with. The technical problems involved are not serious although the techniques may have to be modified for each application because of differing characteristics of the solids and the carriers.

GENERAL REFERENCE

W.S. Gray and P.F. Mason, "Slurry pipelines: what the coal man should know in the planning stage", Coal Age, August 1975, pp. 58-62.

CAPSULE PIPELINES

Technological aspect

1. Present technology

Small diameter capsule pipelines are currently widely used for the short-distance transport of re-usable containers. The usual driving fluid is air and the technique may involve high pressure push, vacuum pull or a combination. The driving fluid may be other gases (such as natural gas) or a liquid (such as oil). Some medium distance capsule pipelines are now in use and others planned. One such pipeline will carry 20 million tons of gravel per year over a distance of seven kilometres (USSR). The capsule may have a shell of a different material and this may be either re-usable or disposable. The usual technical problems encountered are:

(a) when long distances are covered the capsules have to by-pass the intermediate pumping stations;

(b) after shut-down it is difficult to get the capsules moving again;

(c) re-usable capsules are not practical for long distances; disposal capsules may add an unacceptable cost element; and few solids can be treated to form a suitable shell for their own transport.

Economic efficiency: The combined transportation of capsules and a valuable carrier fluid could lead to attractive savings (in terms of money and energy). Capsule pipelines are now economically viable for medium distances.

Importance in ECE region: Large diameter capsule pipelines are not widely used in the ECE region. They have potential for the handling of combined fluid/solid materials over large geographical areas of the ECE region.

2. Methods for reducing environmental impact

Use of environmentally acceptable carriers; recycling of the carriers; back-up safety systems for avoiding accidental releases.

3. General assessment

With the right combination of fluid and solids, capsule pipelines may be very attractive. For long distances the capsules will likely be a homogeneous solid (e.g. purified sulphur) but may be a two-solid capsule: a shell of material composition which has value at the receiving end and a filler of another solid or liquid. A large, stable demand for both the carried fluid and the solid is required.

RELATED TOPICS

- gas pipelines: pp. 12-13.
- slurry pipelines: pp. 68-69

CAPSULE PIPELINES

Environmental aspect

1. Air pollution

   No significant air pollution problem.

2. Water pollution

   Chemical pollution depending on carrier liquid and solubility of the capsule.

   Particulates may become a problem depending on capsule casing and abrasion; not likely to be significant except in case of accidental releases. Possible erosion during construction of the pipeline.

3. Land use

   As for other long-distance pipelines (see pp. 12-13).

4. Solid wastes

   No significant problem if re-usable shell material is used.

5. Noise

   Not serious, except at compressors or vacuum pump stations and during construction of the pipeline.

6. Aesthetic aspect

   No particular problem if the pipeline is buried and the ground revegetated.

7. Others

   If the driving fluid is water, the requirement for large quantities to transport the capsules may be detrimental to local water supplies. Migrations of wild fauna may be affected if pipelines are above the ground.

8. General assessment

   The use of capsule pipelines is unlikely to have serious environmental effects especially if buried pipelines are used. If a non-marketable carrier fluid is used it may have to be purified before being released or recycled.

Chapter Four: Technologies Applicable to Potentially Significant Energy Sources

Contents

GEOTHERMAL POWER PLANTS

Technological aspect

1. Present technology

Method of electrical generation: direct steam and/or hot water from naturally occurring or drilled wells are directed towards turbines. The use of a secondary fluid turbine (chlorofluormethanes or isobutane) can increase the efficiency. Rapid corrosion of equipment is the most important technical problem.

Conversion efficiency: 10-20 per cent.

Economic efficiency: higher economic efficiency than fossil or nuclear fuelled power plants as there is no need for sub-systems (mining, milling, transportation of fuel, etc.). Cost of installation per kW installed capacity: $US 110 (1971). Cost of 1 kWh of produced electricity: 4.86 mills (1971). Drilling wells is expensive.

Importance in ECE region: installed capacity in ECE region in 1975: 982 MWe (Iceland, Italy, United States). Several ECE countries have a good potential (for example, France, Greece, Turkey, USSR).

2. Methods for reducing environmental impact

(i) cooling towers (1.5 to 2.5 more expensive than for conventional power plants);

(ii) brines and used waters are reinjected. Special care should be taken to avoid watercourses and aquifer pollution;

(iii) when feasible, removal of sulphur and boric acid.

3. General assessment

The present very modest utilization of geothermal energy for producing electricity is due to the scarcity of naturally occurring high grade "dry" steam and its location in remote places. Several ECE countries have initiated geothermal survey in order to assess their possible geothermal potential. In spite of the fact that geothermal energy can be considered as a partially renewable source of energy, a careful management of the geothermal "reservoir" is required if the potential is to be maintained over a long period.

RELATED TOPICS

- other types of thermal power plants: pp. 14-31
- new heat transfer media: pp. 50-51
- geothermal space heating: pp. 76-77
- deep geothermal power: pp. 78-79

GEOTHERMAL POWER PLANTS

Environmental aspect

1. Air pollution

Total: 10^4 to 2.10^5 tons/year for a 1,000 MWe power plant but amounts are highly site dependent. The most usual pollutants associated with geothermal steam are SH_2, ammonia, boric acid, fluorides and traces of NO_x, particulates, some radioactive elements such as Ra -222 and Pb -210, as well as hydrogen and methane.

2. Water pollution

Total: 10^5 to 10^8 tons/year for a 1,000 MWe power plant

thermal pollution: 80 to 90 per cent of extracted energy;
chemical pollution: mainly brines (if not reinjected);
radioactive pollution: negligible.

3. Land use

About 20 km^2 for all operations (1,000 MWe)

4. Solid wastes

None.

5. Noise

Steam emerging at high pressure makes a very loud noise (more than 100 db). Important problem, particularly for workers.

6. Aesthetic aspect

Can be an important problem as geothermal energy is generally found in non industrial areas where access roads, drilling rigs, pipes, cooling towers,etc. appear.

7. Others

Potential microseismic effects and local subsidence might appear (both unlikely when using naturally occurring sources).

8. General assessment

There is a definite "natural" impact on the environment due to the occurring nature of surface geothermal energy: flora can particularly be seriously affected; when not used, 100 per cent of the energy is released into the environment producing "natural heat pollution". Using surface geothermal energy should - in theory - improve the situation with respect to chemical pollution if methods for reducing environmental impact are used. Additional problems: noise and landscape disturbance.

GENERAL REFERENCES

"Energy Technology to the Year 2000" Technology Review, MIT, Oct/Nov. 1971, p. 46.

J. Barnea, "Geothermal Power", Scientific American, June 1972, pp. 70-77.

G. Marinelli, "L'Energie Géothermique", La Recherche 5, No. 49, p. 827 (1974).

"Environmental Considerations in Future Energy Growth", Battelle for United States Environmental Protection Agency, April 1973.

GEOTHERMAL SPACE HEATING

Technological aspect

1. Present technology

Method of production: geothermal water is usually pumped from a warm aquifer to a gas separation tank which removes dissolved gases contained in geothermal water when it emerges from the borehole. Warm water is then directly or indirectly used for space heating (houses, schools, factories, greenhouses, etc.). Used water is reinjected into the aquifer through a separate well.

Economic efficiency: a very high economic efficiency is demonstrated when a large source of heat is available. Prospecting costs are of the same order of magnitude as for oil. When an exploitable well has been found, investments are moderate (establishment of heating grid, pumps). The only running costs are maintenance and some energy required for pumping water.

Importance in ECE region: minor. Experience since 1925 in Iceland. District heating in France, Hungary, Iceland and USSR. Greenhouse heating in several ECE countries. Potential findings in practically all ECE countries.

2. Methods for reducing environmental impact

None required if used water is reinjected.

3. General assessment

Large-scale use of geothermal heat for space heating would allow very significant oil and electricity savings (in the Reykjavik district, 200,000 tons/year are saved thanks to geothermal energy) at low costs and without damage to the environment. Geothermal surveys should therefore be promoted.

RELATED TOPICS

- geothermal power plants: pp. 74-75
- deep geothermal power: pp. 78-79
- solar space heating and air conditioning: pp. 80-81

GEOTHERMAL SPACE HEATING

Environmental aspect

1. Air pollution

None, except dissolved gases released at the separation tank.

2. Water pollution

None if used water is reinjected into the original aquifer.

3. Land use

Same order of magnitude as for pipelines.

4. Solid wastes

None.

5. Noise

Very limited (pumps).

6. Aesthetic aspect

No problem. Separation tanks usually look like water tanks.

7. Others

None.

8. General assessment

Naturally occurring geothermal heat produces no significant environmental impact if used for space heating. When available, it is thus highly preferable to oil and fossil fuelled space heating systems or to the use of nuclear electricity.

GENERAL REFERENCES

S.S. Einarsson, "Chauffage urbain grâce aux sources d'eau bouillante", Courrier de l'Unesco, February 1974, pp. 24-25.

D. Spurgeon, "Natural Power for the Third World", New Scientist, 6 December 1973, pp. 694-697.

A. Clot, "La géothermie basse énergie", La Recherche (1977), 8, pp. 213-223.

DEEP GEOTHERMAL ENERGY

Technological aspect

1.1 Present technology

Not operational on an industrial scale at the present time.

1.2 Future technology

Principal objective: Production of electricity.

Principle: Drilling sufficiently deep (8,000 m) to reach high-temperature geothermal fields (300-600°C), with two parallel pipes several hundred metres apart connected within the geothermal field by artificial geological fractures. A fluid (generally water) would be injected into one of the pipes and recovered in the form of superheated steam at the outlet of the other pipe and then carried to turbines. It could operate as a closed-circuit system. Drilling techniques would be similar to those used in the oil industry, in underground nuclear explosions (see pp. 40-41) or based on new methods such as the "subterrene", a device consisting of a very small remote-controlled nuclear reactor, housed in a pointed cylinder approximately two metres in diameter, which would melt the rock, passing through it by gravity and leaving a glass-lined pipe.

Economic efficiency: Not established but likely.

Importance in ECE region: None for the moment. Enormous theoretical potential in a number of ECE countries (100,000 MW in the United States, for example).

2. Methods for reducing environmental impact

Closed-circuit use of the heat-transmission fluid could avoid air and water pollution problems (constraint: salt saturation).

3. General assessment

Neither the technological application nor the economic inability of this method has been proven, but it could apparently supply appreciable amounts of energy without giving rise to major environmental problems.

DEEP GEOTHERMAL ENERGY

Environmental aspect

1. Air Pollution

   Probably fairly similar to geothermal power plants (pp. 74-75) except in the case of closed-circuit use.

2. Water pollution

   Same remark as above.

3. Land use

   Probably of the same order of magnitude as for installations using surface geothermal energy.

4. Solid waste

   None to negligible.

5. Noise

   High-pressure steam; important problem for workers.

6. Aesthetic aspect

   Could be an important problem, depending on the site, drilling rigs, pipes, cooling towers, etc.

7. Others

   Possible microseismic effects and local subsidence.

8. General assessment

   Method favourable from the environmental standpoint, particularly in closed-circuit system operation.

GENERAL REFERENCES

A. Hammond, W. Metz and T. Maugh II, Energy and the Future, American Association for the Advancement of Science, Washington D.C., 1973, pp. 55-60.

"Environmental considerations in future energy growth", Battelle for the United States Environmental Protection Agency, Vol.I (1973), pp. 542-553.

E.S. Robinson and others, "A preliminary study of the nuclear subterrene", Report from the Los Alamos Scientific Laboratory, Los Alamos, New Mexico, United States (1971).

F.H. Harlow and W.F. Pracht, "A theoretical study of geothermal energy extraction", Journal of Geophysical Research, Vol.77, No.5 (1972), pp. 7038-7048.

Otan, Comité sur les défis de la société moderne, "Energie géothermique", (1975), p.162.

SOLAR SPACE HEATING AND AIR CONDITIONING

Technological aspect

1.1 Present technology

Method of production: Direct or indirect solar radiation is absorbed by a black metal sheet enclosed in one or several shallow, glass-enclosed boxes situated on the roof or on the walls of buildings (solar collector). Tubing attached to the black metal sheet carried the heating fluid (usually water) which is stored in a tank connected to a heat exchanger. When needed, water for domestic uses is circulated through the heat exchanger as well as, in advanced and still experimental devices, fluids for heat-operated air-conditioning devices. Often sophisticated systems are assisted by a heat pump.

Economic efficiency: Demonstrated. Relatively high initial capital investment. Very limited maintenance costs (circulating pump). No fuel expenses.

When space heating with normal-sized installations, an additional heat source should be planned in most ECE areas; space heating with solar energy seems competitive with, or even cheaper than, electric heating in most geographical areas. Solar air conditioning seems at present feasible only with large-scale installations (schools, hotels, department stores, etc.).

Importance in ECE region: Insignificant so far except in Cyprus where about one per cent of national energy needs are covered by solar energy. Large potential use. Many ECE governments have decided to encourage and support the use of solar energy particularly for space heating. A few experimental air-conditioning units exist.

1.2 Future technology

Technological research is mostly oriented towards increasing the conversion efficiency of solar energy into heat with the help of "selective surfaces" and to ways and means of storing large quantities of solar heat for all-the-year operation. Solar energy conversion installations capable of heating and cooling large buildings or groups of buildings are in the planning or the demonstration stage.

2. Methods for reducing environmental impact

Improvement of design in order to minimize detrimental aesthetic aspect.

3. General assessment

The wide use of solar energy for space heating seems feasible on a small to medium scale (dwellings, schools, etc.). Obvious advantages are simplicity, absence of distribution network and free energy from renewable source. Its disadvantages are a relatively high initial capital investment and its inability - at the present stage of development - to provide total energy required for heating buildings in many ECE countries. With regard to air conditioning, further research and development is required.

RELATED TOPICS

- geothermal space heating: pp. 76-77
- other applications of solar energy: pp. 82-89

SOLAR SPACE HEATING AND AIR CONDITONING

Environmental aspect

1. Air pollution

   None for solar heating. Possible accidental releases of chlorofluoromethane from certain types of solar-powered air-conditoning units.

2. Water pollution

   None.

3. Land use

   None if solar collectors are placed on roofs or walls of buildings.

4. Solid waste

   None.

5. Noise

   None.

6. Aesthetic aspect

   Minor on dispersed individual houses (collectors). Would be significant in urban areas: majority of south-oriented roofs with very similar slope and appearance. When installed on low, flat-roofed buildings, solar collectors may create an ugly skyline.

7. Others

   None.

8. General assessment

   No serious environmental problems.

GENERAL REFERENCES

- World Meteorological Organization, "Solar Energy", Proceedings of the UNESCO/WMO Symposium, Geneva, 30 August - 3 September 1976; WMO Publication No. 477 (1977).

  Economic and Social Commission for Asia and the Pacific, Proceedings of the Meeting of the Expert Working Group on the Use of Solar and Wind Energy, Bangkok, (ST/ESCAP/7), United Nations publication, Sales No. E.76.II.F.13.

SOLAR STEAM POWER PLANTS

Technological aspect

1.1 Present technology

Presently no industrial application.

1.2 Future technology

Method of production: Utilization of enormous reflectors made up of thousands of sun-tracking mirrors which concentrate sunlight on a steam boiler consisting of a network of pipes through which water or another working fluid circulates. Production of high-pressure steam at temperatures of up to 600° and electricity by means of conventional turbo-generators.

Economic efficiency: France's National Centre for Scientific Research provisionally estimated the construction costs for a 10 MWe solar-powered thermal power station at Ffr.30-80 million (nuclear power station: FFr.20 million). Therefore, not yet competitive with fossil fuel and nuclear power stations. However, this technology seems more attractive than photovoltaic conversion (costs estimated by the same sources at around Ffr. 90,000 per installed kW) (see pp. 86-87).

Importance in ECE region: A few demonstration installations under development or construction in various ECE countries. Large potential use but utilization in the ECE region limited by economic problems, land utilization and climates.

2. Methods for reducing environmental impact

None applicable.

3. General assessment

High initial capital investment but minimum cost of energy production. Intermittent energy production requiring storage facilities such as pumped storage plant. Present technology not competitive with conventional methods. However, the renewable nature of solar energy - in contrast to fossil fuels and uranium - could justify further research and development of this future-oriented technology, particularly for increasing the conversion efficiency, simplifying the system of concentration of the sun's rays and decreasing total contracting expenses up to a level competitive with more traditional electric power stations.

RELATED TOPICS

- fossil fuelled and nuclear power plants: pp. 14-31
- geothermal power plants: pp. 74-75
- photovoltaic conversion: pp. 86-87
- space solar power plants: pp. 88-89

SOLAR STEAM POWER PLANTS

Environmental aspect

1. Air pollution

None

2. Water pollution

Thermal pollution : waste heat if discharged in a water body.
Biological fouling: approximately the same as for conventional electric power stations.
Other types of water pollution: none.

3. Land use

Considerable compared with conventional power stations (30 km^2 for 1,000 MW power station) in addition to the area devoted to conventional equipment: turbo-generators, pumps, cooling towers, electric transmission lines, transformers, etc. Solar power plants would be built in desert or semi-desert areas.

4. Solid waste

None

5. Noise

Less than conventional power stations; not a very serious problem.

6. Aesthetic aspect

Serious problem owing to the "sun tower", the large area occupied by reflectors, water cooling towers and outside installations (high tension lines, transformer sub-stations, etc.).

7. Others

Possible local disturbance of climate and flora if the plant covers a considerable area.

8. General assessment

No serious environmental problems except in regard to land use and aesthetics. Apparently less serious problems than those raised by conventional power stations including gas-fired plants (see pp. 16-17).

GENERAL REFERENCES

"Tower power", New Scientist, 15 April 1976, page 134.

"France's solar boiler", New Scientist, 19 February 1976, page 398.

"Solar energy: Toulouse congress on solar energy", Europe Energy, No. 12, page 15.

A. Hammond, W. Metz, T. Maugh II, Energy and the Future, American Association for the Advancement of Science, Washington D.C., 1973, page 64.

SOLAR DESALINATION

Technological aspect

1.1 Present technology

Method: In most installations, desalination of water takes place in a glass- or plastic-covered basin; the cover is inclined at an angle of 10° to 20°, and the pan for the saline water is painted black to increase the absorption of solar energy. Vapour condenses on the inner surface of the cover and the distilled water flows into troughs on the sides and then into a tank; rainfall can also be collected by means of external troughs.

Economic efficiency: Investment cost: about \$10.5 m^2 per annum; average yield: approximately 1 m^3/m^2; service life: 20 years or more; product water cost: \$0.80 to 1.60 per m^3. By-product: brine. No fuel costs except to pump the water. Seems economical if water requirements are less than 200 m^3 per day. Investment costs are tending to decline, but solar desalination remains more expensive than conventional systems using fossil or nuclear fuels. Main uses: human consumption and animal husbandry.

Importance in ECE region: Technology at the advanced demonstration stage, including semi-industrial scale plants. Now in service in Spain, Greece, USSR and a number of other countries.

1.2 Future technology

(a) Solar ponds storing heat in layers of water of increasing density and permitting the heat to be re-used; high thermodynamic yield but technologically quite complicated.

(b) Combined energy-source systems (for example, cooling water from small electric generators, with the still acting as a cooling tower).

2. Methods for reducing environmental impact

None required.

3. General assessment

The use of solar energy for desalination could be an economic proposition in isolated coastal regions with a suitable climate.

SOLAR DESALINATION

Environmental aspect

1. Air pollution

   None.

2. Water pollution

   None.

3. Land use

   About 1 m^2 per m^3 of water desalinated per annum. In view of the limits on profitability (200 m^3/day), installations are not likely to be larger than 70,000 to 80,000 m^2 (the largest existing installation, at Patmos, Greece, is 8,600 m^2 in size). The problem is likely to remain negligible.

4. Solid waste

   None.

5. Noise

   None.

6. Aesthetic aspect

   Moderate problem. Installations are generally low.

7. Others

   The brine is either disposed of in the sea or evaporated intensively in the open air to extract the salt. Contamination of the desalinated water by bacteria, algae or other micro-organisms.

8. General assessment

   No serious environmental problem.

GENERAL REFERENCES

Solar distillation as a means of meeting small-scale water demands, 1972, (United Nations, publication Sales No.: E.70.II.B.1).

Economic Commission for Europe: Seminar on selected water problems in islands and coastal areas with special regard to desalination and groundwater (Malta, 5-10 June 1978) (WATER/SEM.5/2).

PHOTOVOLTAIC CONVERSION

Technological aspect

1.1 Present technology

Small scale installations of arrays of silicon cells (or of other types of cells) powering very specific equipment such as isolated relay stations.

1.2 Future technology

Method of production: Concentration of solar radiation to arrays of cells by parabolic reflectors tracking the sun by an automatic driving mechanism, by Fresnel lenses or stationary collector. The direct current produced is changed into alternating current and transmitted to distribution networks and/or used for electrolyzing water in order to store hydrogen (pp. 64-65).

Economic efficiency: High but rapidly decreasing capital investments per kW of peak output rendering photovoltaic power stations not yet competitive with fossil fuel or nuclear power plants. In addition, photovoltaic power stations can obviously produce electricity only during sunshine. Therefore it has been calculated, for example for the United States, that a solar installation providing the equivalent to a 1,000 MW fossil fuel or nuclear power station would need a peak output of 2-3,000 MW.

Importance in ECE region: A few demonstration installations under construction. Large potential use in spite of economic, land utilization and climatic constraint problems.

2. Methods for reducing environmental impact

None required.

3. General assessment

The large-scale production of electricity by photovoltaic cells is still not competitive with conventional and nuclear power stations. As the prices of fossil fuels and uranium increase and environmental standards get more stringent, the photovoltaic production of electricity will become more feasible. Research and development is required to increase the conversion efficiency, develop new technologies for large-scale production of solar cells, decrease expenses, store energy, convert electricity into alternative current, etc.

RELATED TOPICS

- conventional power plants: pp. 14-33
- hydrogen production and use: pp. 64-65
- solar steam power plants: pp. 82-83
- space solar power plants: pp. 88-89

PHOTOVOLTAIC CONVERSION

Environmental aspect

1. Air pollution

None.

2. Water pollution

None.

3. Land use

Considerable (10 square km for generating a peak power of 1,000 MW) in comparison with any other known technology except reservoirs of hydro power stations (pp. 32-33). The areas most suited for such installations (deserts) are generally the most remote from large user concentrations. Small scale application would seem much more economical and feasible.

4. Solid waste

None.

5. Noise

None.

6. Aesthetic aspect

Collectors of photovoltaic cells intended for large-scale electricity production could create significant aesthetic problems. No problem with decentralized units.

7. Others

Possible local disturbance of micro-climate. Large areas would probably be closed to people with the exception of the servicing staff. The manufacturing of certain types of solar cells require the use of highly toxic compounds (cadmium, selenium, arsenic, etc.).

8. General assessment

No serious environmental problems, but significant impact on land used. However, attractive technology from the environmental point of view.

GENERAL REFERENCES

World Meteorological Organization, "Solar Energy", Proceedings of the UNESCO/WMO Symposium, Geneva, 30 August - 3 September 1976; WMO publication No. 477 (1977).

B. Chalmers, "The photovoltaic generation of electricity", Scientific American, October 1976.

M. Langues, "Les cellules solaires de demain, "La Recherche", vol. 6, 1975, pp. 870-873.

H. Kelly, "Photovoltaic power systems: a tour through the alternatives", Science, Vol. 199, No. 4329, 1978, pp. 634-643.

SPACE SOLAR POWER PLANTS

Technological aspect

1.1 Present technology

Method: Small arrays of photovoltaic cells (see pp. 86-87) supplying the energy needed to operate artificial satellites.

Economic efficiency: Not proven.

Importance in ECE region: None to marginal.

1.2 Future technology

Method: Very large (64 km^2) geostationary satellites far enough from the Earth to be able to operate around the clock; capacity: 10 to 20,000 MW. The energy produced could be transmitted to Earth after conversion into microwave by ultra-high frequency generators and picked up, regardless of weather conditions, by a receiving antenna of approximately 100 km^2 probably located at sea. Could also be used in space by large-scale industrial satellites.

Economic efficiency: Very large investments.

2. Methods for reducing environmental impact

None required with present technology.

3. General assessment

A method which, if applied on a large scale, seems quite futuristic, at least for the moment, but which could become feasible within one or two decades and possibly resolve some problems (renewable energy, no likelihood of impact on climate, etc.) if used in space.

RELATED TOPICS

- conventional types of power plants: pp. 14-33
- solar steam power plants: pp. 82-83
- photovoltaic conversion: pp. 86-87

SPACE SOLAR POWER PLANTS

Environmental aspect

1. Air pollution

   None.

2. Water pollution

   None.

3. Land use

   High in the case of transmission to Earth of the energy obtained (100 km^2 for 10,000 to 20,000 MWe): none in the case of installations operated only in space.

4. Solid waste

   Probably none.

5. Noise

   None.

6. Aesthetic aspect

   Difficult to estimate: probably not negligible in the case of transmission to Earth of the energy obtained.

7. Others

   Potential effects on living beings accidentally crossing the microwave beam. Photochemical effects on the ozone layer of exhausts from rockets servicing the installation.

8. General assessment

   Artificial transmission of large quantities of solar energy along with that normally received on Earth would eventually, in addition to a number of specific problems, have the same potential climatic disadvantages as conventional methods of producing similar quantities of energy. These disadvantages could, however, be overcome by the *in situ* use of the energy produced.

GENERAL REFERENCES

P. Glaser, *et al.*, *The Journal of Microwave Power*, Special issue on satellite solar power station and microwave transmission to Earth, Vol. 5, No. 4 (1970) 296 p.

G.K. O'Neill "Space colonies: the high frontier", *The Futurist*, Vol. 10, No. 1, 1976, pp. 25-33. This author believes that the design and construction of such colonies are quite feasible on the basis of existing technology and materials. Cost of the first colony for 10,000 persons: approximately $10 billion.

NUCLEAR FUSION

Technological aspect

1.1 Present technology

Studies of envisaged fusion reactors and their engineering problems are undertaken in most countries with fusion research programmes. About 10 per cent of current efforts are devoted to these studies.

1.2 Future technology

Method of production: a mixture of deuterium and tritium is ionized to form a plasma, heated to at least a hundred million degrees centigrade and confined by powerful magnetic fields at that temperature for a sufficiently long period of time to permit a controlled fusion of nuclei to take place. This reaction releases very large amounts of energy which can be used for electricity generation and/or other purposes. Reaction products are principally helium and fast neutrons, the latter being used to produce tritium from a surrounding lithium blanket. Other mixtures are usable as fuel but require still higher temperatures and pose still greater technical problems. Various other means are under investigation for the controlled release of energy using the fusion process, notably the convergence of several powerful laser and charged particle beams.

Economic efficiency: not yet proven

Importance in ECE region: a number of highly sophisticated research installations. It seems unlikely that fusion power could meet a substantial fraction of energy needs before the middle of the next century.

2. Methods for reducing environmental impact

Difficult to evaluate. By utilizing tritium leakage barriers, leakages of tritium could be reduced to very low levels. Careful selection of construction materials, particularly for the core, may minimize the amounts of radioactive materials produced by neutron activation.

3. General assessment

Although the experimental feasibility of controlling a self-sustained nuclear fusion reaction has not yet been demonstrated, much hope has been placed in this technology, as it could provide the release of practically unlimited amounts of energy from abundant basic materials such as deuterium and lithium (and, in more sophisticated versions, from deuterium alone).

RELATED TOPICS

- fission power plants: pp. 22-31
- breeder reactors: pp. 58-59

NUCLEAR FUSION

Environmental aspect

1. Air pollution

   Probably limited to small amounts of tritium.

2. Water pollution

   Unknown. Potential leakages of small quantities of tritiated water.

3. Land use

   Direct land use probably comparable with other types of power plants. Indirect land use would involve modest mining operations, when compared to fission power plants (pp. 22-31).

4. Solid waste

   Activated reactor components with half lives depending on the structural material chosen, thus bearing a high potential for strongly decreasing long-term environmental impact.

5. Noise

   Unknown.

6. Aesthetic aspect

   Unknown but probably comparable to other types of power plants.

7. Others

   No possibility of dangerous nuclear excursion. Decay heat due to activation of structural materials does not constitute a source of danger.

8. General assessment

   Apparently no serious environmental problems but demonstration experiments should have been carried out before assessing this technology.

GENERAL REFERENCES

D.J. Rose, "The prospect of fusion", Technology Review, No. 79, No. 2, December 1976, pp. 21-43

R.L. Hirsch and W.L. Rice, "Nuclear fusion power and the environment", Environmental Conservation 1974, Vol. I, pp. 251-262

BIOMASS ENERGY

Technological aspect

1.1 Present technology

Principle: conversion of the solar energy tapped by photosynthesis and stored in biomass (wood, sugar cane, algae, animal wastes, etc.) into chemicals which can be easily utilized particularly in the domestic sector (heating, cooking, internal combustion engines and others).

Method: anaerobic digestion of organic compounds by the successive action of various types of bacteria. The mixture produced mainly consists of methane (50 - 70 per cent) and CO_2 (25 - 35 per cent). Alcohols, principally methanol and ethanol, are intermediate in the bio-gas production process.

Economic efficiency: not fully demonstrated but seems nearly competitive to more traditional methods. Methanol used as fuel for automobiles, for instance, appears to be between 1 - 1.5 times more expensive than gasoline. Various pilot or demonstration facilities in several ECE countries.

Importance in ECE region: very limited so far but will undoubtedly play an increasing role, particularly in the agriculture sector, which could become self-sufficient in energy. Plays a significant role in other parts of the world: bio-gas fermentors in India, gasoline blended, as an energy saving measure, with methanol produced from sugar-can residuals in Brazil, etc.

1.2 Future technology

"Energy farms" for growing specific plants uniquely because of their ability to easily provide the basic materials for bio-fuel production. These farms could be on land or in off-shore areas.

2. Methods for reducing environmental impact

None are required.

3. General assessment

Simple and efficient method of energy production which could undoubtedly play a growing role in the ECE area.

RELATED TOPICS

- gas fired power plants: pp. 16-17
- various applications of solar energy: pp. 80-89

BIOMASS ENERGY

Environmental aspect

1. Air pollution

   The combustion of methane produces air pollutants comparable to those resulting from burning natural gas (see pp. 16-17). Automobiles fuelled with methanol emit, for slightly diminished performances (acceleration) and for the same mileage, about six to ten times less carbon monoxide, nitrogen oxides and unburned hydrocarbons than those consuming gasoline and require no lead additive as an anti-knock agent.

2. Water pollution

   No apparent problem. The growth of aquatic plants and algae as raw materials for the production of bio-fuels in polluted water bodies might help in solving eutrophication problems.

3. Land use

   Negligible, except in the case of energy farms.

4. Solid waste

   Residuals from the fermentation process are generally regarded as excellent fertilizers.

5. Noise

   No apparent problem.

6. Aesthetic aspect

   No serious problem.

7. Others

   Explosion hazards with tinkered bio-gas tanks. Soil depletion with inappropriate farming. Possible erosion.

8. General assessment

   The production of methane and alcohol from biomass wastes is probably the most favourable renewable energy technology from the environmental viewpoint as it provides not only a "clean energy" supply but also a potential tool for combating water pollution problems, managing organic wastes from municipal facilities, agriculture, food industries, etc., and decreasing the air pollution load from internal combustion engines.

GENERAL REFERENCES

UNITAR "Microbial energy conversion", Proceedings from a seminar, Göttingen 4 - 8 October 1976, published by E. Goltze KG, Göttingen.

Economic and Social Commission for Asia and the Pacific "Report of the workshop on bio-gas technology and utilization", Bangkok, 1975, document E/CN.11/IHT/L.18.

A.L. Hammond, "Photosynthetic solar energy: rediscovering biomass fuels" Science, Vol. 197 (1977) pp. 745-746.

E.S. Lipinsky, "Fuels from Biomass: Integration with Food and Materials Systems", Science, Vol. 199, No. 4329, 1978, pp. 644-651.

W.C. Saverny and D.C. Cruzan, "Methane recovery from chicken manure digestion" Journal of the Water Pollution Control Federation, Vol. 44, No. 12 (1972), pp. 2349-2354.

TIDAL ENERGY

Technological aspect

1.1 Present technology

Principle: Utilization of the tidal difference in the level of the sea. Minimum difference: 4 m. The same water can, in principle, pass through the turbines twice (flood and ebb tides).

Economic efficiency: High investment costs (over $800 per kW installed). Variable hourly output.

Importance in ECE region: Minor. Two plants now in operation: La Rance, in France, and Kislaya Guba, in USSR.

1.2 Future technology

Principle: Large barrages, creation of complex reservoirs, connexion with the hydroelectric system (storage lakes).

Economic efficiency: Very high investment costs, more stable hourly production; possible increase in local tourism.

Importance in ECE region: Will continue to be minor, despite major projects such as those in the Bay of Fundy and Ungava Bay (Canada), the Bristol Channel (United Kingdom), Cook Inlet, Alaska (United States of America), Mezen and the Sea of Okhotsk (USSR).

2. Methods for reducing environmental impact

Apparently none required.

3. General assessment

Attractive method of energy conversion, but practical application is and will remain very limited.

RELATED TOPIC

- hydropower plant: pp. 32-33.

TIDAL ENERGY

Environmental aspect

1. Air pollution

   None

2. Water pollution

   None from the plant itself, but stagnation of sewer water could give rise to serious problems. Possible accidental releases of oil from turbines and transformers.

3. Land use

   Practically none, but very large marine areas required.

4. Solid wastes

   None.

5. Noise

   Negligible.

6. Aesthetic aspect

   May be regarded as unimportant or extremely important, depending on personal opinions, beauty of the site, etc.

7. Others

   Apparently no effect on fish passing through the turbines (low rotation speed). It may be added that areas with very strong tides are not of particular value or interest from an ecological point of view, and that some attenuation of natural tidal movements might even prove beneficial. Silting. Effect on marine communities.

8. General assessment

   Method of producing energy with virtually no impact on the environment.

GENERAL REFERENCES

International Conference on the Utilization of Tidal Power, Halifax, Nova Scotia, Canada, Plenum Press, New York-London (1972).

T. Shaw, "Tidal power and the environment", New Scientist, 23 October 1975, pp. 202-206.

G. Godin, "The power potential of Ungava Bay and its hinterland" Water Power, May 1974, pp. 167-171.

"The Kislaya Guba tidal electric power station", published by Energia, Moscow, 1972.

R.H. Clark, "Re-assessing the feasibility of Fundy tidal power", Water Power and Dam Construction, 1978, Vol. 30, No. 6, pp. 35-41.

WIND ENERGY

Technological aspect

1.1 Present Technology

Principle: Direct utilizations (sailing and water pumping) or conversion of the wind's kinetic energy into electricity. In the latter case the output of a rotor installation increases according to the cube of the wind velocity and the square of the blade length; limitations: strength of materials, production of energy is intermittent.
Methods: (a) Horizontal-axis rotor, blades facing the wind and slow rotation speed. Approximately 40 per cent of the wind's energy is converted into electricity (theoretical maximum: 59 per cent). The largest existing installation appears to have a capacity of 2 MWe (Tvind, Denmark).

(b) Vertical-axis turbine, with two or three rigid or flexible blades turning at high speed. Wind direction is immaterial. Conversion efficiency: approximately 35 per cent. Particularly suitable for local applications.

Economic efficiency: Investment cost from $200 kW (Tvind) to over $2,000 per installed kW. No fuel costs and low maintenance cost.
Importance in ECE region: Formerly considerable, particularly in the Netherlands and in Denmark. Very limited at the present time: usually regions that are not easily accessible (Arctic, islands, etc.), small settlements, etc. Research now under way in most ECE countries.

1.2 Future technology

Methods: Refinement of present techniques, increasing the size of installations by the use of stronger materials, (notably those used in aeronautical engineering), connexion to energy storage systems in order to offset the intermittent production of this form of energy.
Economic efficiency: Could probably become competitive with conventional forms of energy conversion.
Importance in ECE region: Potentially significant, but seems to depend primarily on research and development in this field.

2. Methods for reducing environmental impact

Apparently none required.

3. General assessment

Potentially an important source of energy, involving technological principles that should be further developed, particularly as regards storage methods. High conversion yield possible because of the high quality of the energy used (kinetic energy).

RELATED TOPICS

- conventional power plants: pp. 14-33
- hydrogen production and use: pp. 64-65
- battery storage of electricity: pp. 66-67
- applications of solar energy: pp. 80-89

WIND ENERGY

Environmental aspect

1. Air pollution

   None.

2. Water pollution

   None.

3. Land use

   Negligible.

4. Solid waste

   None.

5. Noise

   Practically none.

6. Aesthetic aspect

   Large-scale installations could disfigure sites considerably.

7. Others

   Metal blades interfere with some radio waves (particularly television).

8. General assessment

   Wind power offers many advantages from an environmental standpoint, its sole drawback being the disfigurement of sites by large-scale installations.

GENERAL REFERENCES

D.M. Simmons, "Wind power", Energy Technology Review, No. 6, 1975, Noyes Data Co., London.

T.W. Black, "Megawatts from the wind", Power Engineering, March 1976, pp. 64-68.

J. Bockris, "Possible means of large-scale use of wind as a source of energy" Environmental Conservation, 1975, Vol. 2, No. 4, pp. 283-288.

D. Hinrichsen and P. Cawood, "Fresh breeze for Denmark's windmills", New Scientist, 10 June 1976, pp. 567-570.

A. Bruckner, "Taking power off the wind", New Scientist, 28 March 1974, pp. 812-814.

C.A. Brown, "La turbine à vent intéresse l'Amérique elle-même" Coopération, published by A.C.I.D., (Ottawa), August 1974, pp. 14-19.

N. Wade, "Windmills: the resurrection of an ancient technology", Science, 1974, Vol. 184, pp. 1055-1058.

UTILIZATION OF OCEAN TEMPERATURE GRADIENTS

Technological aspect

1. Present technology

Method of production: semi-submerged offshore power plants using natural differences in ocean temperatures, particularly in tropical or warm sea current areas. Surface waters of about 27°C are pumped through a heat exchanger where a "working" fluid with a low boiling point, such as ammonia, propane or chlorofluoromethanes, is vaporized by the heat from the water. The vapour expands through the turbine, which powers a generator. The expanded vapour at low pressure is then cooled in a condenser where deep (300-1,000 m) and cold (5°C) ocean water circulates and is then recycled to the heat exchanger. Maximum possible efficiency is about 5 per cent but actual efficiency would be only 2 or 3 per cent.

Economic efficiency: not demonstrated. Capital investments evaluated at about $2,100/kW for a 100 MWe plant.

Importance in ECE region: a few very small scale pilot installations of little economic importance.

1.2 Future technology

Refinement of present techniques, particularly low-pressure turbines and "working" fluids. Increased size of installations to 100 MWe or more. Long-distance sub-marine transmission of large quantities of electricity.

2. Method for reducing environmental impact

None seems applicable.

3. General assessment

The use of ocean temperature gradients for the production of electricity calls upon relatively simple techniques (except during construction). It presents the advantages over most other renewable energy technologies of requiring no energy storage facilities.

RELATED TOPICS

- conventional power plants: pp. 14-33
- new heat transfer media: pp. 50-51
- geothermal power plants: pp. 74-75
- solar steam power plants: pp. 82-83
- photovoltaic conversion: pp. 86-87
- space solar power plants: pp. 88-89
- tidal energy: pp. 94-95
- wind energy: pp. 96-97

UTILIZATION OF OCEAN TEMPERATURE GRADIENTS

Environmental aspect

1. Air pollution

   None foreseeable except for leakages of heat transfer media (see pp. 50-51).

2. Water pollution

   Anti-fouling agents.

3. Land use

   Irrelevant.

4. Solid waste

   None.

5. Noise

   Unknown.

6. Aesthetic aspect

   Minor due to the semi-submerged character of the installation.

7. Others

   Very large discharges of unusually cold water near the sea surface may affect tourism, marine communities and fisheries and may lead to an increase in absorption of solar energy of the ocean surface layers (albedo) and consequently to some potential meteorological effects. Potential development of mariculture due to the high nutrient content of deep-sea waters. Possible effects on ozone layer in case of repeated releases of chlorofluoromethanes. Large obstacle to navigation.

8. General assessment

   An attractive energy technology from the environmental point of view. It seems to be the only large-scale and centralized application of renewable energy technology for the production of electricity which seems to have no significant environmental impact.

GENERAL REFERENCES

W.D. Metz, "Ocean thermal energy: the biggest gamble in solar power", *Science* 1977, Vol. 198, No. 4313, pp. 178-180.

G. Haber, "Solar power from the oceans", *New Scientist*, 10 March 1977, pp. 576-578.

O. Roels and coll., "Organic production potential of artificial upwelling marine culture", in UNITAR "Microbial Energy Conversion", Proceedings of a seminar, Göttingen, 4-8 October 1976, published by E. Goltze KG, Göttingen.

WAVE ENERGY

Technological aspect

1.1 Present technology

Principle: wind energy stored in sea waves as mechanical oscillations is converted by various technical means such as pistons into high pressure air or water which is then used for generating electricity. Other applications, notably sea water desalination, and other principles are also under examination. A power storage device is required in view of the intermittent and variable character of waves.

Economic efficiency: not demonstrated

Importance in ECE region: small scale prototypes of no economic importance.

1.2 Future technology

Several devices using wave energy would pump sea-water atop a cliff area into large storage reservoirs. Water would flow back to sea through a classical hydroelectric plant. This system would be flexible and could be used in conjunction with other renewable sources of energy particularly wind power.

2. Methods for reducing environmental impact

Apparently none required.

3. General assessment

Suitable technology for local application or for achieving very special tasks in coastal areas, which deserves research and development support.

RELATED TOPICS

- applications of solar energy: pp. 80-89
- tidal energy: pp. 94-95
- wind energy: pp. 96-97
- utilization of ocean temperature gradients: pp. 98-99

WAVE ENERGY

Environmental aspect

1. Air pollution

   None

2. Water pollution

   None except if anti-fouling agents are used

3. Land use

   Irrelevant except if storage reservoirs are used

4. Solid waste

   None

5. Noise

   Apparently none

6. Aesthetic aspect

   Minor except for installations in cliff areas

7. Others

   Power transmission to onshore facilities. Possible interference with navigation, tourism and fisheries.

8. General assessment

   Pollution-free technology which is attractive from the environmental point of view.

GENERAL REFERENCES

"Wave motion can be used to tap wind energy", Energy International, April 1975, pp. 19-20

"Resonance makes wave power tunable", New Scientist, 24 March 1977, p. 702

Index

This index is not intended to be comprehensive: only key words which might be of interest to the environmentalist dealing with energy questions have been listed. Furthermore, such words as "air pollution", "noise" or "energy", which are of a very general nature and relate to practically every topic, have been omitted.

Key to symbols

* asterisk	:	indicates that the key word is the topic of a two-page examination
-- long dash	:	replaces the key word
( ) parentheses	:	indications relating to the key word
[ ] square brackets	:	indications to the reader

* * *

O

P